APOLLO, CHALLENGER, COLUMBIA

The Decline of the Space Program

A Study in Organizational Communication

Phillip K. Tompkins
University of Colorado at Boulder, Emeritus

With the Assistance of
Emily V. Tompkins

Roxbury Publishing Company
Los Angeles, California

Library of Congress Cataloging-in-Publication Data

Tompkins, Phillip K.
 Apollo, Challenger, Columbia—the decline of the space program: a study
in organizational communication / by Phillip K. Tompkins with the assistance
of Emily V. Tompkins.
      p. cm.
 Includes bibliographical references and index.
 ISBN 1-931719-32-2
 1. Challenger (Spacecraft)—Accidents. 2. Columbia (Spacecraft)—Accidents. 3. Project Apollo (U.S.) 4. Space vehicle accidents—United States—Case
studies. 5. Communication in organizations—United States—Case studies.
6. Corporate organizations—United States—Case studies. 7. Organizational
behavior—United States—Case studies. I. Tompkins, Emily V. II. Title.
TL867.T65 2005
363.12'4'0973—dc22

                                                            2003022468
                                                            CIP

APOLLO, CHALLENGER, COLUMBIA—THE DECLINE OF THE SPACE
PROGRAM: A STUDY IN ORGANIZATIONAL COMMUNICATION

Publisher: Claude Teweles
Managing Editor: Dawn VanDercreek
Production Editor: Jim Ballinger
Production Assistant: Nina Hickey
Copy Editor: Arlyne Lazerson
Proofreader: Anton Diether
Cover Design: Marnie Kenney
Typography: Abe Hendin

Printed on acid-free paper in the United States of America. This book meets the
standards of recycling of the Environmental Protection Agency.

ISBN 1-931719-32-2

                ROXBURY PUBLISHING COMPANY
                        P.O. Box 491044
                Los Angeles, California 90049-9044
             Voice: (310) 473-3312 • Fax: (310) 473-4490
                   Email: roxbury@roxbury.net
                   Website: www.roxbury.net

Left to right: the author, Walter Weisman, and W. Charles Redding (ca 1982).

Dedicated to the Memory of
W. Charles Redding
Walter Wiesman

Contents

foam

Preface

Perhaps the best way to begin an explanation of how I'm qualified to write this book is to describe my relationship with the two men to whom it is dedicated. I was fortunate to have been recruited by W. Charles Redding to work under him in pursuit of my doctorate at Purdue University. Redding had almost single-handedly created a new field of study, now called organizational communication. As my adviser he created a plan of study for my graduate work, including classes in economics, the sociology of organizations, inferential statistics, criticism, rhetorical and communication theory.

I was fortunate to be in the first generation of Redding's graduate students in what is now a large and growing field. Our aim then was to understand complex organizations as communication systems. We took as our assumption that organizations cannot function, indeed cannot exist, without communication. We came to believe over time that organizations are *constituted by communication.*

An organization comes into existence when one or more persons recruit others to join them in pursuit of purposes and objectives. Following Chester Barnard (1938), a successful executive and management theorist, we believed that the first function of an executive is to establish and maintain a system of communication, a structure, linking all members, clients, and customers. We studied upward- and downward-directed channels and media, feedback loops, and an enduring concept Redding labeled the "communication satisfaction" of organizational members. Superior-subordinate communication was a hot topic at Purdue and remains today the most heavily researched topic in the field.

Redding's pioneering seminars were enriched by his experiences as a corporate consultant with a major aerospace company. He also testified as an expert witness and consultant to General Electric in an extremely important case heard by the National Labor Relations Board. He taught us, however, that an organizational consultant could do more damage than an inept brain surgeon. He also taught us to avoid the "pro-management bias" of certain business schools. We were to treat all members of an organization as equal in importance, if not in status. In an attempt to demonstrate the even-handedness of his program, he persuaded me to do my Ph.D. dissertation on an international labor union. In the process of my research he guided my attempt to create and measure the concept of semantic-information distance in human hierarchies, a topic that applies to NASA today.

Semantic-information distance is the degree of understanding-misunderstanding among people at different levels of an organization, as well as the degree of knowledge-ignorance of ideas crucial to the organization's purpose and rules. We developed a method of measuring semantic-information distance that combined quantitative and qualitative data. After I completed my doctorate, I was hired as an assistant professor at Purdue and was given the task of teaching the first undergraduate course in organizational communication.

After three years—in 1965—I moved on to Wayne State University in Detroit as an associate professor; I taught graduate seminars and workshops for management and labor unions in the Industrial Relations Center. In early 1967 I received a phone call from a total stranger in Huntsville, Alabama, with a German accent. He introduced himself as Walter Wiesman, the Coordinator of Internal Communication at the George C. Marshall Space Flight Center, then the largest field center in the National Aeronautics and Space Administration (NASA).

Wiesman explained later that he was one of the Germans brought from the German space program to the United States in Project Paperclip. They had surrendered to the U.S. Army and were carefully screened before being brought to Fort Bliss, Texas. They tested their uprated V-2 rockets at White Sands, New Mexico, for scientific and military purposes. Later they were transferred to the U.S. Army's Redstone Arsenal in Huntsville, Alabama. They did the research and development and fabrication of the Army's arsenal of intercontinental ballistic missiles during the Cold War with the Soviet Union.

Wiesman was the youngest of the 120 Germans brought over and the only one without a scientific or engineering education. He had learned at the German space center in Peenemünde on the Baltic the importance of communication. So had his boss, the legendary rocket wizard, Dr. Wernher von Braun. When many of the Germans were transferred, over the Army's protest, to NASA in 1960, they assumed many of the leadership positions in the new George C. Marshall Space Flight Center (MSFC). Von Braun was the first director of the space center; he had discovered on his own the crucial importance of organizational communication as the Technical Director in Peenemünde.

Von Braun encouraged Wiesman to develop a program in organizational communication for the education of the Marshall Center employees. He was also encouraged to bring in experts to conduct research at the space center. Wiesman had joined several professional organizations devoted to the study of communication and had met W. Charles Redding. Redding gave Wiesman my name as a potential consultant to NASA. Would I, asked Wiesman, be willing to spend the summer of 1967 as a Faculty Consultant to NASA at the Marshall Center? I would be treated as a civil servant, a GS-13, and would have to hurry to get in my detailed application for employment and a security clearance.

I jumped at the chance. NASA's field centers brought in professors in scientific and engineering disciplines from universities around the country to serve as Summer Faculty Consultants. They most often conducted and directed research projects. Someone told me I was the first or one of the first "soft" scientists brought into this program.

Wiesman was willing to place a bet on me despite my relative youth—I was 33 at the time—and then explained what my duties would be. I would work eight hours a day (sometimes longer hours) reviewing and evaluating Wiesman's program in internal communication, including his five-year plan, conducting research at the center, and supervising research done by doctoral students from Purdue at the Marshall Center. Wiesman had also organized a national conference on organizational communication in August of 1967 for interested representatives from aerospace contractors and other businesses and people from government agencies and the academic world. I was expected to help coordinate the conference and deliver the central paper, or lecture.

The conference was the first of its kind and was a great success. I had by August learned that engineers produced documents with what I thought were ironic titles: "state-of-the-art" papers. And so I gave my paper the title, "Organizational Communication: A State-of-the-Art Review." I organized and summarized the findings from the first approximately 100 empirical research studies done in the field. (Today no one could possibly cover all the research in a single lecture because of the explosive growth of the field.) NASA published the paper as part of the proceedings and later issued it as a separate monograph.

I was, however, learning more than I was teaching. Wiesman put me through a thorough orientation program of visits to the laboratories, the museum—I read von Braun's correspondence with Albert Schweitzer about the possibility of afterlife for humans—and the test facilities. I watched a test of the mighty F-1 rocket engine, described below, one of five on the Saturn V rocket that would launch the Apollo space capsule. The atmosphere at the Marshall Center that summer was electric with excitement and anticipation. They had done the research and development of the moon rocket there, and the first test flight was scheduled later that year—1967. Its first flight to take astronauts to the moon was due to take place in 1969, only two years in the future. I got caught up in the excitement like everybody else.

I also attended the Fifth Annual Summer Lecture Series in Aerospace Science and Engineering sponsored by NASA, the Marshall Center, and the U.S. Army Missile Command. A certificate proclaiming P. K. Tompkins had successfully completed the lecture series (20 hours) hangs proudly in my study today. I am forever grateful there wasn't a final exam! The only lecture I can remember was on the "Third Body Problem." It was about a mathematical riddle of what would happen if a third body entered the gravitational field of two other bodies. I'm told the problem has since been solved by a supercomputer; the reason I remember the lecture is because the mathematician delivered the entire lecture standing on one leg, the other crossed at the knee, all the while tempting the law of gravity.

A big event was my first interview with Director von Braun. He was 55 years old then and larger than life. I had seen Hollywood's film biography of him. Born in Wirsitz, Germany, in 1912, he received his Ph.D. in physics at the age of 22 in 1934 from the University of Berlin; for reasons of military security his dissertation title was somewhat misleading: "About Combustion Tests," but it contained a theoretical investigation, supported by experiments, of a rocket engine. He immediately went to work for the German Ordnance Department and by 1934 he and his group had launched successfully two A-2 liquid-fueled rockets; they reached altitudes of about one and a half miles. He became the technical director of Rocket Center Peenemünde in 1937, developing and launching the V-2 rockets during World War II. When the Soviet Army was nearing the Rocket Center at the end of the war, von Braun led the majority of his rocket specialists from the eastern part of Germany to Bavaria. There they surrendered to Western Allies and were taken to Fort Bliss, Texas, and eventually to Huntsville, Alabama.

In Chapter Four of this book I summarize that first interview, explaining his philosophy of organizational communication. We decided in that meeting that I would spend the rest of the summer conducting a diagnostic study of the Marshall Center as a communication system, finding out what worked well and what didn't. Von Braun wanted to discover whether there were problems in the system connecting the 7,200 employees at the Marshall Center with the total of 200,000 people in the agency and in contractor organizations working on the Apollo Project.

First Data Point

During the rest of the summer I was given complete access to the organization, interviewing in depth about 55 top rocket scientists and managers at the Marshall Center, the Michoud Assembly Plant near New Orleans, Louisiana, and the Mississippi Test Facility in Hancock County, Mississippi. I discovered much that worked, ingenious communication techniques detailed in the fourth chapter of this book. I also found some problems—blockages and barriers—and worked to come up with recommendations for overcoming them. I also worked on action items for von Braun. He was curious, for example, about the Saturn V Control Center, a management facility to coordinate all activities having to do with the moon rocket. NASA headquarters had built it and thought it would be a major contribution. (I found that it was not working well; briefings were thereafter rescheduled in another room von Braun preferred.)

At the end of the summer I briefed von Braun in his office. He was delighted to hear about some powerful effects of a reporting system he had created; he was concerned, however, with some of the serious problems I had unearthed. At the end he jumped up from the conference table to get a calendar from his desk. He looked up the dates of his staff and board meetings in the fall, and we selected one on which I would return from Detroit to brief the center's top management. I did so in the room where von Braun preferred to get briefings, the very room where I had sat in on briefings for him on such complex topics as the possible trajectories for a spacecraft to take in a trip from the Earth to Mars. (An engineer sitting next to me leaned over and said, "You can't hardly get there from here.")

I still have an audiotape of the two-hour briefing I gave the top rocket scientists of the world. They interrupted me, challenged and questioned me, some with German accents. They were clearly worried about some of the problems, illustrating an openness of communication about problems I've rarely seen in other organizations. There was also considerable laughter, sometimes at their expense. They invited me to come back the next year, 1968, again as a Summer Faculty Consultant in Organizational Communication.

Second Data Point

In the winter and spring Wiesman kept me informed about how my recommendations were being implemented. When I returned in 1968, von Braun and Wiesman both had projects for me to pursue. NASA headquarters had ordered a deep Reduction in Force, or RIF in the inevitable acronym, which would mean layoffs for many Marshall employees now that the Saturn V's first test flight had been a success. The Marshall Center would have fewer employees to work on more projects, a diversification in which they would do the research and development for what was called the Apollo Applications Program, including projects such as Skylab and the Apollo Telescope Mount.

Von Braun asked me to help him with a needed reorganization of the center. He also had a list of action items for me to research. Again I was given complete access to documents (many of which were classified) and to interviewees. Coming up with a new organizational scheme—which von Braun regarded as the formal communication system—was the most complex problem I ever faced. But I was helped by the people who would have to implement it and live within it. After I briefed von Braun about my preferred scheme, he whipped out a copy he had been working on and compared it to mine: "We're close," he said.

I had now become an insider with two data points on the space center, 1967 and 1968. The experience had an enormous impact on my subsequent teaching and research. I had noticed during those two summers how NASA employees worked long hours at half the money the contractor personnel made. They nonetheless had the deepest sense of identification with the organization I've ever seen, the deepest sense of commitment to their projects I've seen. I felt it myself. That realization led me to engage in a career-long research program into the phenomenon I call—without notable creativity—organizational identification.

I also watched how the engineers and scientists, out of their identification with the organization and its projects, created working groups and teams not specified by the organization chart of this government bureaucracy. They often functioned without a formal leader, caught up in "automatic responsibility" for problems they perceived. They shared control of the group's deliberations. Later I would come to call this "concertive" control, a group of people working together in concert to solve organizational problems. I'm probably best known in my own field for my theoretical and empirical work on organizational identification and concertive control—with help from associates at three different universities. The seeds for these programs were found in the Marshall Space Flight Center during my experience of what is now regarded as the glory years of NASA.

I was invited to return as a Summer Faculty Consultant in 1969, but I had become a professor of communication at Kent State University in 1968; Kent State gave me a travel grant allowing me to spend part of the summer doing research on the great Irish novelist, James Joyce, in Dublin and London. I expected to return to the Marshall Center in the summer of 1970 to continue my research. On May 1, 1970, however, a tragedy occurred on my campus. Four students were killed and nine wounded by the Ohio National Guard during a demonstration against the U.S. incursion into Cambodia. I was asked to serve that summer on a commission to investigate the causes of the tragedy and chair a task force on communication. Out of that came a book: *Communication Crisis at Kent State* (Tompkins and Anderson 1971). In addition, we reversed the consulting process by bringing Walter Wiesman to the campus as a speaker and consultant to our administration and faculty about the tragedy.

In 1971 I became a professor and department chair at the State University of New York at Albany and organized a research project with colleagues on organizational identification. In addition, I published two articles about my NASA research in a journal, *Communication Monographs* (Tompkins 1977; 1978). Wiesman read these articles in manuscript form, and reprints of the publications are in the archives of the space center today. Wiesman and I were close friends by then (and remained so until his recent death), and I continued to be a proud NASA watcher over the years. Then came word about the tragic *Challenger* disaster of 1986. By then I had returned to my alma mater, Purdue, as a professor and associate dean of Liberal Arts. Purdue had a close connection with NASA; astronauts Gus Grissom and Neil Armstrong, the first man on the moon, were both graduates of Purdue.

The Marshall Space Flight Center (MSFC) had responsibilities for the space shuttle despite having been forced by the Nixon administration and NASA headquarters to accept a complicated and compromised design involving liquid and solid rocket fuels. (Von Braun once told me we shouldn't ever put astronauts on top of a rocket with solid fuels because such fuels couldn't be tested adequately and couldn't be turned off once they had been ignited.)

Third Data Point

As I followed the news reports about the *Challenger* accident, I couldn't believe my eyes and ears. The media reported some communication practices at the Marshall Center that wouldn't have been tolerated during the Apollo days. Ronald Reagan appointed a presidential commission, the Rogers Commission, headed by former Secretary of State William P. Rogers. As I read the commission's report I came to the conclusion that NASA had forgotten much of what we knew and learned in the 1960s. I formulated a theory of organizational forgetting and tested it by visiting the Marshall Center for two days of interviews in January of 1990.

The results confirmed my hypothesis. Von Braun had resigned in 1970 to take a job in industry and died in 1977. Wiesman had long been retired as well. The workforce had been cut in half. Too few people were trying to manage too many projects. Few of my inter-viewees could remember the communication philosophy and tech-niques of the Apollo era. I agreed with the Rogers Commission that, ironically, miscommunication had been a factor leading up to the tragic accident. I planned to write journal articles about this research, my third "data point," but, as I explain in the Introduction to this book, Claude Teweles persuaded me to write an academic book for his publishing company: *Organizational Communication Im-peratives: Lessons of the Space Program* (Tompkins 1993). Someone from NASA who read it e-mailed me about its implications for an agency program called "lessons learned," but the interaction was superficial and short.

Fourth Data Point

I continued to study NASA. In the summer of 1987 I studied NASA's Aviation Safety Reporting System (ASRS) and visited the facilities of its contractor, the Battelle Memorial Institute's office in Mountain View, California. They allowed me to interrogate their huge data base of reports about near accidents among airplanes. The ASRS is an interorganizational communication system for the entire aviation industry. Input is provided by controllers and pilots and other observers. Those who write reports about the "incidents" are given immunity from prosecution. The reports are read by ana-lysts—former pilots and air traffic controllers—who store them in the computer by categories. By the time I visited the computer in 1987 they had accumulated 70,000 reports involving 50,000 inci-dents (there may be more than one report per incident). All of these had been processed and cross-checked by a second analyst without a single breach of the promise of anonymity and immunity.

The system worked effectively, even brilliantly, to prevent avia-tion accidents. As of 1987, NASA had issued more than 835 alert bulletins, time-critical notices about hazards such as overgrown trees blocking approaches to an airport runway and faulty naviga-tion lights and runway markings. My questions put to the data base had to do with human factors. I was not surprised to discover that between 60 and 70 percent of the reports in the data base included a reference to obstacles to human communication (sometimes coded as problems of "information transfer") as a contributing factor in the incidents considered to be serious threats to aviation safety. In a third of these miscommunication problems somebody—perhaps a controller—forgot to say something to someone else—a pilot, for example. In another third, incomplete or inaccurate information was communicated by one to another, leading to incorrect decisions about controlling and piloting airplanes. One-eighth of the incidents involved communicating in a tardy manner. In one-tenth of the reports, the receiver either didn't get the word or misunderstood the message.

I was invited to present my findings to a conference in Boulder, Colorado, on man-made risks sponsored by the National Science Foundation. It was published as a book chapter edited by the organizer of the conference (Tompkins 1990). A different version was published under the title "Organizational Communication and Technological Risks" (Tompkins 1991). The paper was released to the press, and I suddenly found myself on radio talk shows from coast to coast. An airline pilot on the way to O'Hare called in to a talk show in Chicago to confirm my conclusions. The study was summarized in *Popular Mechanics.* An interview with a Denver television station was picked up by CNN and I heard from friends around the country about it and from one colleague who saw it while in Australia. It seems that most people naïvely underestimate the importance of communication and miscommunication, and it is newsworthy when evidence to the contrary is introduced. In that book chapter I reviewed my research with NASA in the first three data points, hoping that lessons from the Apollo Program and NASA's own Aviation Safety Reporting Program would help reduce technological risks. This was the fourth "data point" in my relationship with NASA.

I continued to follow other studies of the *Challenger* accident. In 1997, for example, I wrote an essay-review about all of the books about the tragedy for an international journal, *Organization* (Tompkins, Heppard, and Melville 1998). I gave lectures trying to reconcile the conflicting interpretations of the causes of the accident, a project carried forward in the present volume.

Fifth Data Point

Then the sky fell. *Columbia* broke up on February 1, 2003; seven astronauts were killed. Although again depressed, I decided not to pursue the case. And yet I couldn't stop reading about the accident and the investigation of it. In April I was invited to give a lecture via telephone to students in an organizational communication class at Michigan State University. As I prepared for it, I heard what Dr. Sally Ride called an "echo" of the *Challenger* accident. The analysis in my book about the first shuttle accident seemed to fit or apply to the evidence in the second shuttle accident.

When I agreed to write this book, it became my fifth "data point" on one organization. This time, however, it would be different. I had to concentrate on NASA-as-a-whole, not a field center. The evidence this time would come from newspaper accounts and the Report of the Columbia Accident Investigation Board (CAIB). I was no longer an insider. In the end that became a positive factor because it gave me the needed perspective given by distance.

I saw no reason to interview anyone this time because I saw no evidentiary holes in the voluminous newspaper articles, the CAIB Report, and in November of 2003 a long article about *Columbia* that appeared in the *Atlantic Monthly*. The author had gained access to CAIB and NASA. His work confirmed that additional interviews by me were unnecessary. My contribution would be to give pictures of NASA over time so that the reader could follow the dramatic decline of a once-proud institution. In addition, I was able to accept the evidence presented by CAIB and the press and make a somewhat deeper interpretation of the mechanisms underlying what is now the conventional explanation. Finally, I have placed NASA in the context of other contemporary American organizations and institutions in an attempt to find a more general explanation in Chapter Eight. In that chapter I present eleven of what I call "communication transgressions" that are common to most of them and that appear to be correlated to individual and organizational failures. Ethical issues emerge in the narration and analysis, from start to finish. Only a careful reader can determine whether these new interpretations have traction in helping us understand our organizational environment. Whatever the decision of that careful reader—James Joyce thought the perfect reader is an insomniac—I remain a Professor Emeritus of Communication and Comparative Literature. *Emeritus* is a Latin word for more time to play golf and to write books. ✦

Introduction and Acknowledgments

The first person to suggest this book was a former colleague at the University of Colorado at Boulder: David Noller. His main argument was that I would have three "data points" on one organization over a 35-year stretch, an unusual perspective. (It turns out that there are five of those data points—five periods in which I made an intense study of NASA.) Although I considered his proposal briefly, there were a couple of other writing projects with higher priorities.

I was ultimately persuaded by Claude Teweles, the publisher, to write this book, the second time he's done that to me. *Organizational Communication Imperatives: Lessons of the Space Program* (1993) was also his idea. My plan was to publish my research on the *Challenger* accident as articles in academic journals, just as I had done in 1977 and 1978 when articles of mine about NASA and the Apollo Program appeared in *Communication Monographs.*

I don't regret being easy to persuade, even though writing this book put a strain on me and my family, but this time I was inclined not to write. I was already trying to write an ethnographic book about a homeless shelter in Denver where I have worked as a volunteer for five years. After the *Columbia* accident I got inquiries from professors and students reading my book about *Challenger;* they were curious as to whether I saw any similarities between the two shuttle accidents. In fact, I agreed to give a lecture via telephone to Professor Vernon Miller's class in organizational communication at Michigan State University on April 24, 2003. Having enjoyed the interaction with the students immensely, I considered revising, mainly updating, the existing book, but the thought of all that research was daunting.

When Claude called, he said he wanted a new book, not a revision, so at first I resisted his efforts to persuade me. He did get me to agree to think about it. Where I was and what I was doing is now lost to me, but there was a moment in which I saw the book in five chapter divisions in my mind. There are now more than that, a change forced on me mainly by the massive body of information I gathered about the second shuttle accident. Important to me was the fact that I could see a keynote address I had given in late January of 2003 to the Rocky Mountain Communication Association as having relevance to the final chapter; for 62 minutes I spoke to 115 professors and graduate students about the "Crisis of U.S. Organizations and Institutions." A few days later, when I learned that the *Columbia* and her crew were lost, I knew NASA had to be added to my list.

Another source of motivation came from my own persuasive attempt to recruit my daughter, Emily V. Tompkins, to be my research and editorial assistant. She has been in the book business for some time and can work the internet better and write better than I. She agreed to help and soon had me buried with reprints from the Archives of the *New York Times*. When the CAIB report came out . . . there I go again, NASA's fever for alphabet soup has again infected me . . . when the report of the Columbia Accident Investigation Board (CAIB) came out, I asked several people to help me a get copy of it. The first copy I received was from Emily. She also made the book better by asking questions a lay reader would have and by improving my prose with more felicitous phrases. For these reasons I changed my mind, said yes to Claude, and wrote this book.

I again speak in the first person singular. It seems more natural and colloquial to me. And I think it was Henry David Thoreau who said it's always the first person speaking whether we acknowledge it or not. The contraction in the previous sentence is a sign that the style of the book is an attempt to be informal for the reasons given in the last paragraph of this introduction.

Reference to the CAIB prompts me to say that the first three chapters were completed—except for minor editing—before the CAIB report was issued on Tuesday, August 26, 2003. Knowing from experience with the *Challenger* accident that the book couldn't be completed until the official report was released, I proceeded with the method of the police procedural, the genre I discovered on a trip to Sweden in 1977 in the work by Sjowall and Wahloo, books such as *The Man Who Went Up In Smoke, The Fire Engine That Disappeared,* and *The Man On The Balcony.* Their books begin with the fact of a crime—that much is known. The narrative unfolds as the police gather evidence; as they gather evidence, they develop hunches and track down false leads. My version of this fictional method was to begin not with a crime but with the fact of the *Columbia* accident. As new information is collected, it is presented to the reader, day by day, week by week. As the evidence is obtained, it is presented as the reporters presented what they and NASA were finding and learning; the organizational past of NASA is presented in flashbacks.

Quite early in the investigation a reporter referred in print to Sherlock Holmes, making me realize that Sir Arthur Conan Doyle's method was similar to the police procedural; that inspired a brief discussion of *The Hound of the Baskervilles* and Sherlock Holmes' method of deduction. Reporters wrote of the "mystery" and "clues," and a NASA briefer spoke of NASA's "detective work." A key figure in the launch described NASA's graphic method of reverse engineering, the fault tree analysis, a cognitive method analogous in a formalized way to the methods of a mystery and police procedural.

In all of these methods we begin with what is known, whether it is a murder or a shuttle accident, then branch out into possible explanations. Hypotheses have a tendency to multiply, increasing in a way positively correlated with the complexity of the case. The detective or the engineer must be able to discriminate among hypotheses, rejecting those that do not help explain the evidence. The most costly part of the search is the inductive process of bringing facts, evidence, or data to bear on hypotheses in either a positive or negative way. Some hypotheses can be rejected, some can't. In the end our deductions are qualified as probable causes; in the moral realm, however, there are some certainties.

These ideas about presentation emerged during the search itself, from reading news conferences and reporters' articles. I give a tip of the hat to the reporters who covered this case. They were persistent from the first days in asking questions NASA did not enjoy. Indeed, certain NASA officials now appear to have been in denial about possible causes, but the reporters kept pressing them for answers. The *New York Times* had suffered a serious blow when they discovered a reporter had broken a commandment of journalism, but in this case the newspaper regained its credibility by aggressive and accurate reporting, by raising important questions, and by pressing NASA constantly for answers. They quickly corrected minor errata, including mistaken claims given by NASA officials. The news reports and government documents at times provided opposing viewpoints.

In the narrative chapters about the *Columbia* accident I have avoided the scholarly impedimenta of footnotes, sources, dates, and page numbers. Instead, the names of reporters are used in the text, sometimes with their datelines, including abbreviations. All the articles are from online archives and don't have the same page numbers they would in the printed version; each article begins with page one, page two, and so on. The reader with access to the internet can therefore easily track down any of the articles used; in the meantime the reporters get credit. In other chapters I have used a style of references common to the sciences, physical and social, the use of parenthetical and shorthand pointers to the fuller citations in the list of references.

As mentioned earlier, I have adopted a more informal style than in the past. Retiring from the university five years ago loosened my style somewhat. In addition, my experiences in NASA and the space program during the Glory Years taught me they were primarily oral cultures; the work got done in oral discourse, complete with acronyms and contractions. It seemed more natural to me at times to approximate the less formal style of speech. It does occur to me it may also be an unconscious desire to return to that glorious culture of the past.

Claude Teweles also persuaded a number of people, all but one in the field of communication, to read the first draft of the manuscript. They made many useful suggestions I've tried to follow: Connie A. Bullis, University of Utah; Tim Hegstrom, San Jose State University; Stephen Johnson (Space Studies), University of North Dakota; Greg Larson, University of Montana; Larry Long, Illinois State University; May Meares, Washington State University; Vernon Miller, Michigan State University; and John Oetzel, University of New Mexico.

I want to thank Gregory Desilet, David Noller, and Elaine Tompkins for reading the book in manuscript form and for making some excellent suggestions; I also want to thank Tom Jensen and Laura Bloss for suggestions they made in conversation. Finally, thanks again to ET for her love and support. ✦

Chapter One

The *Columbia* Accident

More than a dozen people read this book in manuscript form, most of them professors who are concerned greatly about how their students learn. Most of them advised me that it was their experience that a lot of students don't bother to read Prefaces and Introductions, preferring instead to get to the action immediately by starting with the first chapter. Such students, they said, and perhaps other readers as well, would find it difficult to get oriented to the unusual organization of this book. Some of the professors themselves had trouble getting oriented; they needed a forecast, a kind of roadmap of the book. So at the risk of repetition, I will restate some of what is in the Introduction and Preface to help the reader follow my format.

Forecast

I think of this book as a story, a nonfiction novel, the outcome of which is known to most if not all of my readers. That is similar to a crime novel in which newspapers and other media report a murder. The suspense is created by curiosity about *why* and *how* the murder was committed. Although there was no murder, no crime, in the case of the *Columbia* accident, there was a dreadful accident with loss of life. How did it happen? Why did it happen? Rather than summarize what the final investigators discovered, I have chosen to present the information as it came in immediately after the disaster. I relied on press reports of NASA news conferences, first-hand accounts of the debris falling from the sky, and interviews with observers published in the Archives of the *New York Times* and other newspapers. I also consulted NASA's website. All of this is important to an understanding of how to analyze organizational catastrophes.

By coincidence, a friend and neighbor, Gregory Williams, was at NASA's Kennedy Space Center on February 1, 2003, waiting to watch the shuttle *Columbia* land. I begin this chapter with an interview I later conducted with him; from there I summarize the news and "evidence" as it became known and reported, a method used in the mystery genre known as a "police procedural," in which the reader learns along with the police and detectives. We see what hypotheses are developed and how the incoming data tend to confirm or disconfirm them. We also see how television was not able to cover the story the way they did the *Challenger* accident in 1986. This chapter describes what evidence was gathered and reported during the day after the accident and ends with a list of possible technical causes of the accident.

Chapter Two continues this narrative process for the week following in a frenzied attempt at sense making. Members of the press realized early on that they were dealing with a mystery—rather, two mysteries. The first mystery is the technical or technological one; what went wrong with the shuttle? The second mystery is about communication, organization, management; what, if anything, did NASA do wrong to create the technical failure? Reporters following the investigation talked about clues and even mentioned Sherlock Holmes, the famous English detective created by Sir Arthur Conan Doyle. The reader learns, for example, of a tension between the official investigating commission, the Columbia Accident Investigation Board (CAIB), and the agency being investigated, the National Aeronautics and Space Administration (NASA). This is an important lesson to learn: *Large organizations invariably try to protect and defend themselves from outsiders.* Journalists who were not seeking to protect and defend the organization began to compare this accident with the earlier shuttle accident. They began to recall that NASA had been less than open in discussing the *Challenger* accident 17 years earlier—a communication transgression; some said NASA had tried to cover it up. This introduces a theme that will be developed throughout the book: the important but infrequently examined topic of the *ethics of communication.* I introduce my own tentative hypothesis about how language could have led to a misunderstanding that contributed to the accident. The narrative of this chapter concludes at the end of the next day, February 2, 2003.

Chapter Three takes a break from the narrative format to introduce a brief history of NASA and concepts from the field of communication such as organizational culture, the Ideal Managerial Climate, and superior-subordinate communication, which will become central to the investigation and analysis of the *Columbia* accident. That chapter presents a detailed description of the original "strong" technical culture. Chapter Four continues this analysis by concentrating on the biggest field center of NASA, the Marshall Space Flight Center (MSFC), during the glory days of the space program: the Apollo Project, which sent Americans to the moon and safely brought them back home to Earth. That chapter reports some of the research I did as a Summer Faculty Consultant at the Marshall Center in the 1960s on communication-as-culture, linking it to the success of the organization.

Chapter Five looks back at the *Challenger* accident of 1986. Not long after the *Columbia* accident a lot of people, including me, saw some similarities between the two shuttle accidents. It is necessary, therefore, to move backward in time to understand how various scholars and a Presidential Commission analyzed the first accident. My analysis is presented along with several others so that the reader can decide whether my synthesis of them makes the best sense. Organizational communication is one of the best, if not the ultimate, perspective on this national tragedy; returning to it helps one understand, by analogy, the second shuttle accident.

Chapter Six returns to the narrative of the unfolding mysteries of *Columbia* with the press coverage until the release of the Columbia Accident Investigation Board (CAIB) Report in late August. Quoted in that chapter are e-mail messages among NASA employees relevant to the accident that were released to the press. They show that some members of NASA suffered from misunderstanding, from a form of miscommunication. Some central NASA figures were reassigned from their posts so that they would not appear to be investigating themselves. CAIB began to issue recommendations.

Chapter Seven is a close reading of the CAIB Report. It came out seven months after the accident, on Tuesday, August 26, 2003. I consider it the final chapter in the police procedural, in which the mystery is solved. By means of a technique called fault-tree analysis, certain hypotheses, called "branches" of the tree, are eliminated one by one. The remaining branch was deemed to be the technical cause of the accident. But there is a remaining mystery—what were the organizational causes of NASA's failure? Communication and culture, more or less synonymous, are examined in careful detail. Using CAIB's evidence and observations, I provide my own analysis. Mysteries solved.

These first seven chapters are an in-depth case study, the analysis of one organization over nearly forty years. It is my belief that by getting to the heart of one organization, one gets to the heart of all organizations. But rather than assume that is so, Chapter Eight looks at other organizations that were in decline at the same time NASA had its failure. I present evidence in the cases of Enron, the Catholic Church, and other troubled organizations, showing that their difficulties are similar to those of NASA. These organizations suffered from forms of unethical communication practices, miscommunication, and other mistakes made by NASA.

Chapter Nine tries to wrap up the book. It explains how *Spiderman*—the film, that is—fits into the larger picture of the book. It also asks what a student should do in preparation for an organizational life. It will be hard to avoid a career that doesn't involve organizational communication. What practices should one look out for? In what situations should the individual resist the demands of an organization? How can one recognize when the cultural force of an organization—a mighty power—must be disobeyed or at least ignored?

Now, I suggest you get into position to dive into the mysteries. If you have difficulty with any passages of the book, you can reach me via e-mail at <Tompkinp@Spot.Colorado.Edu>. Enjoy.

The Mystery

Gregory Williams, 51, was born in Detroit, Michigan, and received a degree in civil engineering from Michigan State University in 1975. He now works as a project engineer in Denver, Colorado, specializing in the renovation and restoration of high-rise buildings. He describes himself as a Black American; his success in a technical career led him to create a Foundation, The Real Me, by which, in cooperation with the Denver Public Schools, he identifies minority students and girls with scientific or technical aptitudes and links them via e-mail with mentors at the state's private and public colleges, universities, and business organizations. Once a year he rounds up his charges, buses them to meet their mentors on campus and in other organizations. In addition, the pupils meet with admissions advisers so that they can prepare for moving on to higher education and a technical career. His cousin from California was going to be spending a couple of weeks in Orlando, Florida, at a seminar in late January and early February of 2003, and asked Gregory to join him for a few days of vacation.

Williams had never spent any time in Florida other than in airports, so he accepted the offer, arriving in Orlando Thursday night, January 30. The next day, Friday, January 31, he and his cousin read the local newspapers over breakfast to find out what was happening and learned that the *Columbia* space shuttle would be landing the next morning at Kennedy Space Center (KSC), a drive of about 90 minutes from their hotel in Orlando. As an engineer, Williams was interested in the space-shuttle-as-machine, as well as the process of space flight. The cousins agreed to make the trip; his cousin's co-worker from California, also attending the seminar, asked to join them.

They arrived at the Visitors Center at KSC the next morning at about 8:15 Eastern Standard Time. There were only about 150 or 200 people, Williams told me (our formal interview was conducted on Monday, August 11, 2003). That was because landings are not as spectacular as launches. Local residents pay little attention to them—except for the sonic boom produced by the shuttle's descent. The Center had movies about the space program and souvenirs, but the three men walked outside to look over the facilities. The Center was placed in such a way that the spacecraft would come down directly over it on its flight path toward the landing strip.

The first announcement over the public address system came at approximately 8:45; the content was, as Williams remembered six months later, that *Columbia* had reached California at about 20,000 miles per hour and at about 200,000 feet. The next announcement, five minutes later, said the craft was flying at about 18,000 miles per hour at about the same altitude over Texas and should be landing in about 15 minutes. That was the last official announcement Williams and the others heard.

After 15 minutes passed, everyone began to peer into the skies. Birds and a surveillance plane gave false hopes. After 20 minutes of silence the crowd became "inquisitive," Williams said, curious about the delay. People had pulled out their cell phones to share the moment with friends and relatives. Some of them were told that the shuttle would not be landing, that there was "a problem or an incident." These people quickly spread the word among the 200 people or so at the scene. "Somber" was the word Williams used to describe that moment. It was also, as he later added, "suspenseful."

People rushed to the souvenir shop. The three men joined them. Their thinking was that this was a moment of History—with a capital H—and that they needed a memento of the momentous occasion. Williams bought a cap embellished with a shuttle emblem and the names of the seven astronauts. "In fact," said Williams, "I intended to bring the cap with me today as a gift for you, but I forgot it." After they bought their souvenirs, the three men walked back outdoors to see the huge NASA buses, like "Greyhounds," said Williams, head down to the landing strip to pick up the VIPs, the relatives and friends of the seven astronauts. Williams could only speculate as to how difficult that scene would be—notifying the loved ones—but took it as a strong "indicator" that something bad had happened. They thought the scene might become "hectic," so they walked to their car.

"As we drove away the television crews were coming in the opposite direction like a flock of vultures." They turned on the radio for news, "mesmerized and in shock." Williams said, "My vacation was null and void," and he wondered, "What made us come here today?" He paused and said: "Things happen for a reason—even if you can't figure them out—and you have to live with them." He paused to reflect and then added that the experience "was like a bad dream. It wiped out our whole day. I didn't want to do anything but listen to the radio and watch TV. Then sporadically we would start talking to each other about it." He also said they did some drinking while they talked.

Williams has a placid demeanor and an easy smile but is not otherwise easy to read in regard to his emotions; still, he seemed to be somewhat irritated when he complained that NASA was so "secretive." The only information they got at KSC was via cell phones. "They [NASA] didn't have a news conference for several hours," complained Williams.

It seemed as if our interview was about to die of a lack of energy; suddenly Williams remembered something he thought was important: "We began calling people like crazy from the hotel room. You're part of history so you want to tell people; it sucks you into the moment."

Q: Who did you call?

A: The four most important people in my life.

I had another question. In my reading about the accident I had learned that if NASA had realized there was a problem with the left wing they could have brought *Columbia* in at a different angle, one in which the right wing would have handled most of the load and the heat, temperatures between 2,000 and 3,000 degrees Fahrenheit. Williams said there were two other possibilities as well. The second was to send up a second shuttle to earth's orbit, rescue the astronauts, and destroy the *Columbia*. The third was to send *Columbia* to the International Space Station where it could wait for another spacecraft to rescue them.

(But you would have had to have known that the left wing had been seriously damaged.)

Debris Rains on Nacogdoches, Texas

While Gregory Williams and the others at KSC were experiencing shock, David M. Halbfinger and Richard A. Oppel, Jr., were filing an article with the dateline Nacogdoches, Tex, Feb. 1, with this title: "LOSS OF THE SHUTTLE: ON THE GROUND; First the Air Shook With Sound, And Then Debris Rained Down." The article would appear in the *New York Times* the next day. The article begins with four one-sentence paragraphs, the first three of which are:

> It sounded like a freight train, like a tornado—like rolling thunder—and then a gigantic boom.

> It fell from the sky in six-inch chunks and seven-foot sections of steel, ceramics, circuit boards and who-knows what.

> It tore holes in cedar rooftops, scorched front lawns, ripped a streetlight from its pole and littered the parking lot behind the Masonic hall downtown.

Miraculously, despite the pieces of the space shuttle *Columbia* that fell to the ground on hundreds of square miles of eastern Texas and western Louisiana, no one was injured—on the ground that is. A hospital worker in Hemphill, Texas, was horrified to find the charred torso and skull of an astronaut near some pieces of debris on a rural road. Human remains were also found in Sabine County, Texas.

NASA and local officials warned people to stay away from the debris because of the dangerous and deadly brews of chemicals, even carcinogenic substances, on the shuttle. Law enforcement officers from all levels of government tried to find and cordon off the largest pieces. At the center of Nacogdoches itself, a crowd gathered around a large piece of roped-off metal, placing bouquets, praying in circles, and some treating the event as an "alien crash." And there were unconfirmed reports that people were offering bits of debris as souvenirs on eBay. The FBI was to take jurisdiction; they warned that scavengers would be prosecuted.

John Anderson, 59, said, "We heard this low-frequency, high-energy sound, an enormous release of energy, sort of a ragged boom." He found over 70 pieces of debris on his 14-acre lot of grass and trees. The first piece that landed on his porch got his attention; he remembered that a shuttle was scheduled to land that day and feared the worst. He ran in to turn on the TV. "We had the TV on, and by that time they were reporting there had been no communication. But we already knew."

Mr. Anderson knew much more than the people on television or those watching television: "One remote television image, small globes of light trailing plumes of smoke across the sky, said everything that could be said in the minutes and hours after NASA reported a mysterious emergency with the space shuttle *Columbia*'s return to earth." So wrote Alessandra Stanley in the *New York Times* on February 2, 2003. Her article about television coverage that Saturday is one of nearly two dozen articles about the tragedy appearing in that newspaper the day after the accident. Stanley gives credit to CBS's Dan Rather for being the first anchor on the news set, going live just one hour after the shuttle disintegrated.

Saturdays are usually slow news days; journalists and reporters and correspondents like to take a weekend off as much as anyone. Rather got to work before the others, but with few facts he and the other correspondents were at a loss for words, depending on memory, experience, and their ability to improvise. Taking a call from a man who said he was an eyewitness, Rather asked the caller to describe what had landed on his property. The man said it was the teeth of an astronaut, adding that the broadcaster was an "idiot."

CBS cut the caller off quickly; Rather explained there was no way to avoid crank calls, and as an afterthought said, "I am an idiot, but that is beside the point." Mr. Rather, a native Texan, also spoke of Nacogdoches, Texas, as "an old Indian trading post" located behind "what is called the Pine Tree Curtain," before reporting that a considerable amount of debris had landed there.

Brian Williams appeared on NBC an hour later. Bill Blakemore took the desk at ABC until Peter Jennings went on the air at about noon. Blakemore was quicker than others to raise such important questions as the effect of a loss of a shuttle on the International Space Station, dependent as it is on the space shuttles for supplies for the two astronauts and one cosmonaut aboard. (Aaron Brown, anchor of CNN for eighteen months, was playing golf in the Bob Hope Celebrity Tournament in Palm Springs, California, and he didn't make it to work until the day after the accident.) Tom Brokaw of NBC broke off his vacation in the Virgin Islands and was able to anchor the evening news on Saturday. CBS announced at 10:57 a.m. that flags were flying at half-staff at NASA Headquarters. The networks were understandably reluctant to say that the shuttle and the seven astronauts were lost. They did, however, raise the possibility that what NASA was calling a "contingency" might be an act of terrorism. The Fox network interviewed Dan Gillerman, Israeli delegate to the United Nations, about the Israeli air force officer on board the *Columbia*; Gillerman said Colonel Ilan Ramon was the child of a Holocaust survivor and the leader of an attack on a nuclear reactor in Iraq in 1981. It was hard work for the television commentators because they had so little hard news and so few images to pass along to viewers. They knew that NASA had lost all signals, all communication from *Columbia*, and that debris was scattered over eastern Texas and western Louisiana. Later we would learn the last voice message from the spaceship was "Roger, uh. . . ."

The Internet

John Schwartz contributed a contrasting picture of the television coverage with his article in the *New York Times* titled: "LOSS OF THE SHUTTLE: THE INTERNET, A Wealth of Information Online." Not only did it provide people with information, it also provided them with a medium in which to disseminate it:

That was nowhere more clear than on the high-tech community known as Slashdot, at <www.slashdot.org>, where members posted more than 1,100 messages by 5 p.m. that included links to NASA pages, first-person accounts of hearing or seeing the breakup, the text of Ronald Reagan's 1986 elegy [sic] to the *Challenger* astronauts, arguments over the future of space travel, and the usual exchange of insults that crop up in any online discussion.

Within hours of the disaster, a man named Don Drake had downloaded images of the orange trail of debris across Texas from the radar website of NOAA, the National Oceanic and Atmospheric Administration. He combined and animated these images so they moved every second or so for anyone on the internet to see. He was interviewed by Schwartz by phone and said that this was something that the conventional media do not do very well, that they don't have the variety of technical talent to be found on the internet. Using the jargon of journalists, he spoke of the thousands of technical "stringers" out there online. According to another of Schwartz's sources, one could see the story of the accident develop by reading through the postings in the order in which they appeared.

The President Speaks

A silent confirmation that the astronauts were lost came when the White House lowered its flag to half-staff at about noon. President George W. Bush, informed at Camp David of the disaster by his Chief of Staff, Andrew H. Card, Jr., sped down the mountain and through the Maryland countryside to the White House in a motorcade. He looked "drawn and stricken" as he spoke on television at 2:00 p.m. from the Cabinet Room: "The *Columbia* is lost. . . . There are no survivors." Echoing President Reagan's remarks after the loss of *Challenger,* he said the space program would continue, and to honor the five men and two women, he added: "The same creator who names the stars also knows the names of the seven souls we mourn today. . . . We can pray they are safely home."

A NASA News Conference

A news conference was held in the afternoon by Ron D. Dittemore, Shuttle Program Manager, and James M. Heflin, Jr., NASA Mission Operations chief flight director at the Johnson Space Center (JSC) in Houston. The *New York Times* published excerpts from it the next day. They expressed their shock at losing seven family members. It was a great day to land in the Florida area, they said, because the weather was cooperating with the landing. They had had nothing but positive signs until a few minutes before 8:00 a.m. Central Standard Time. Mr. Dittemore said the first sign was the loss of temperature sensors in the left wing, followed by the loss of tire pressure on the left main gear, and then indications of excessive structural heating: "I have to caution you that we cannot yet say what caused the loss of *Columbia*. It's still very early in our investigation and it's going to take us some time to work through the evidence, the analysis, and clearly understand what the cause was."

Mr. Heflin took his turn after Dittemore to say that when they noticed the abnormalities they sent an alert to the astronauts, an alert to check their displays. He then had difficulty articulating whether the crew knew that something was wrong: "I can't—I've asked a couple of people. I haven't heard the tapes myself. I'm not sure what they said at the time, but they were acknowledging, we believe, that indication that they'd seen." Then they lost all data from the vehicle at around 8:00 a.m. Central Standard Time; the craft was at 207,135 feet, traveling at a Mach of about 18.3. "And the flight control team during this time—again, we lost the data and that's when we clearly began to know that we had a bad day."

A reporter at the press conference stated that he had heard reports that some debris hit the wing during the launch and then asked if that was true and a cause of concern. Mr. Dittemore acknowledged that a piece of foam used as insulation on the external fuel tank (ET) hit somewhere on the left wing; they weren't sure where it hit, he added. Given their experience with tile on the wing, he said, "It was judged that that event did not represent a safety concern. And so the technical community got together and across the country looked at it, and judged that to be acceptable."

Houston, Texas, Loses Heroes

The press conference was held at the Lyndon Baines Johnson Space Center (JSC) in Houston, about 200 miles south and a bit west of Nacogdoches. Houston is also the home and training center of the astronauts. In an article for the *New York Times* with the dateline Houston, Feb. 1, reporters expressed a NASA-wide theme that would be repeated many times by many people—death in the family. The city was in mourning because Houstonians were proud to say they lived in the same city or same neighborhood with an astronaut's family. They were heroes, even though the current crop lives a quieter life style than the earlier generations, who held keg parties the night before a launch. Perhaps this indicates a cultural shift in the astronaut corps, a change from a cowboy culture and the "right stuff" into, well, more like an accountant's culture.

At Frenchie's Italian Restaurant, off NASA Road 1, pictures of the crew hang on the wall near the cash register. There is a photograph of a recent birthday party held for the pilot of the lost flight, Commander William McCool. The owner of the restaurant, Frankie Camera, was sad for several reasons, one of which was they died over Texas, "so close to home." Reporters Bragg and Yardley described the disbelief and resignation and numbness and sadness in Houston: a funereal atmosphere. They also described the press conference at JSC, how the two administrators spoke with cracking voices; they wrote about the speakers' "precise language about failed sensors and damaged tiles on the left wing of the shuttle." Mr. Heflin was also quoted as articulating another long-time NASA theme: when something breaks, we fix it.

Reactions at the Cape

An article with a dateline from Cape Canaveral, Florida, Feb. 1, by Dana Canedy had this headline the next day: "LOSS OF THE SHUTTLE: CAPE CANAVERAL; Keenly Felt Grief in a State Entwined with the Space Program." The reporter interviewed tourists at KSC, local residents, and officials. A woman from Alexandria, Virginia, was quoted as saying about the *Columbia* crew: "You hope that somehow these lives were spared." Tourists were no longer in the mood for recreation, some deciding to cut their vacations short. Much like Gregory Williams, people felt they had lost their vacation; some visitors said they were "stunned," and a woman from West Virginia said, "You feel kind of hollow inside."

Canedy's article moves from the effect on tourists to description of the unique relationship between Florida and the space program. After the explosion of *Challenger* on January 28, 1986, the Florida legislature remembered the seven-person crew with a special license plate, one that turned out to be among the most popular ever.

Senator Bill Nelson, a Democrat, took a ride on the *Columbia* itself in 1986 when he was a congressman representing Brevard County, in which the space center is located. Public officials said they were shocked and saddened over the loss of *Columbia*.

At the runway, where NASA officials, reporters, and relatives were awaiting the spacecraft, "panic" took over as the shuttle failed to appear. The Administrator of NASA, Sean O'Keefe, in office only a year, "fighting back tears, addressed the news media a short time later: 'This is indeed a tragic day for the NASA family, for the families of the astronauts, and, likewise, tragic for the nation.' "

*   *   *

Unlike the *Challenger* accident, which appeared on live television with a huge nationwide audience of schoolchildren, in this case people were more likely to hear the news in anxious telephone calls and radio broadcasts and cell phones and the internet. Dean E. Murphy's article in the *New York Times* on February 2 has the subtitle: "Sorrow, Memories of *Challenger* and a Will to Move Ahead," based on the responses of citizens around the country. Some citizens told Murphy they were already apprehensive about a possible war with Iraq, memories of 9/11, and the *Challenger* accident. Others linked *Columbia*'s fate with fears about the accident's effect on the economy. Many thought it might be a terrorist attack. Sympathy for Israelis was expressed by many. A pediatrician in Grosse Pointe, Michigan, hoped the loss might cool the "fiery rhetoric" toward Iraq. A store manager in Buchan, Michigan, said she watched President Bush's speech on television with tears in her eyes.

Readers of the Sunday *New York Times* and other newspapers were able to sense the outline of the event the following day, February 2, 2003. In an overview article, one of the many pieces in the paper that day about the accident, David E. Sanger in a dateline from Washington, D.C., explained that *Columbia* had broken up during re-entry into the atmosphere of the earth, killing all seven astronauts and raining fiery debris across Texas and Louisiana. The loss, he said, would revive the long-term debate about the space program in Congress and would renew questions about management problems at the agency.

The Crew and Its Mission

Best known among the astronauts was Ilan Ramon, 48. He was selected as an astronaut candidate in 1997 following a science agreement between President Bill Clinton and Shimon Peres, then the Israeli foreign minister. The son and grandson of Holocaust survivors, Ramon felt he represented Israelis and all Jews as the first of his ethnic identity to fly in space. On the shuttle he was in charge of an Israeli project to gauge the effect of dust storms on climate. He carried a special symbol onto the spaceship: a small Torah scroll used in the bar mitzvah of the project's principal investigator, Dr. Joachim Joseph, nearly 60 years earlier while he was in a Nazi concentration camp. Security surrounding the flight was extraordinarily tight because of Ramon's status as a national hero in Israel. Experts on terrorism, wrote Sanger, thought it was highly unlikely that *Columbia* had been hit because it flew at such a high altitude, beyond the reach of conventional weapons. Mr. Bush called Ariel Sharon, Prime Minister of Israel, and the two friends "grieved together," as did their two nations. Ironically, we would later learn that the shuttle broke up near a small town in east Texas named Palestine.

The flight was commanded by an Air Force Colonel, Rick D. Husband, 45, and the pilot was a Navy Commander, William C. McCool, 41. The scientific payload was coordinated by Michael P. Anderson, 43, a Lieutenant Colonel in the Air Force, the third African American to die in the service of the space program; Dr. Kalpana Chawla, 41, a woman born in a province of India where girls are often aborted, who had a doctorate in engineering from the University of Colorado and had become an American citizen; and two Navy doctors, Captain David M. Brown, 46, and Navy Commander Laurel Salton Clark, 41. The mission of this flight was unusual, being completely dedicated to scientific experiments, some ninety of them, including eleven from schools worldwide and, as mentioned, the Israeli study of dust storms. A more typical mission for the shuttle is the transport of people, equipment, and supplies to the International Space Station and the support of military operations, but *Columbia*'s cargo bay could carry less mass than other shuttles because, as the first one built, it had less-advanced structural materials. Like all shuttles, *Columbia* got her name from a ship, in this case a Boston-based sloop that discovered the Columbia River on the

Left to right: David Brown, Rick Husband, Laurel Clark, Kalpana Chawla, Michael Anderson, William McCool, Ilan Ramon (CAIB Report, Vol. I, p. 29).

coast of Oregon and was the first American vessel to circumnavigate the world. This was the 113th shuttle mission, the 28th for *Columbia*, the oldest of the Orbiters. That means, of course, that the rate of catastrophic failure for the shuttle is two in 113.

Early theories of accidents are often disproved, wrote Sanger, but he couldn't resist recalling that a few days earlier NASA had revealed that a piece of foam insulation had hit the left wing during the launch. A similar incident had been observed in a previous launch but without inflicting any major damage. All theories would be reviewed by NASA itself and by an independent board chaired by Admiral Harold W. Gehman, Ret., who was one of the two chairmen of a commission that investigated the terrorist attack on the U.S. destroyer *Cole*.

Then Sanger made the inevitable comparison to the *Challenger* accident, observing that whatever had happened to *Columbia*, it was different from what caused the first shuttle disaster. The infamous O-rings of the *Challenger* were part of the solid-fuel boosters; the *Columbia* problem seemed to be centered in the left wing. The first disaster happened on ascent, the second on descent, on re-entry into the earth's atmosphere where it is subjected to temperatures in excess of 2,000 degrees. *(It seemed to me, however, that both shuttles might have been doomed during the launch.)*

Turning to the problem of the International Space Station, Sanger stated that the two Americans and one Russian aboard were scheduled to be picked up and returned to Earth by *Atlantis*, set for a launch on March 1. The station, he wrote, had a large stock of water, food, and other supplies, enough to support the three for several months. The Russian Space Agency was also scheduled to send a robotic cargo ship, *Progress*, to the station, but the building materials to be used to expand the station could only be hauled by one of the three remaining shuttles.

The latter part of Sanger's overview article skipped, in journalistic style, from topic to topic, mentioning in random order how communication was cut—"It's as if someone just cut the wire," quoting Mr. Dittemore. NASA had declared a "mission contingency [anomaly]," when anyone on the ground could see that the spacecraft had broken up. There was none of the drama of *Challenger* because the shuttle was not in full view of the television cameras. Debris was scattered over hundreds of square miles of Texas and Louisiana; in Hemphill, Texas, a driver "came across what appeared to be parts of the remains of an astronaut." The space program had lost the political import it had during the Cold War. There was another Colonel Ramon story:

> [H]e had little room to take personal items on the flight, but he did lift off with a piece from the Holocaust-era: a small black-and-white drawing called "Moon Landscape" that he had borrowed from the Yad Vashem Art Museum in Israel. The drawing, by Peter Ginz, a 14-year-old Jewish boy killed at Aushchwitz in 1944, was a picture by a child who dreamed of faraway places and sketched what he thought the Earth would look like from the mountains of the moon. This morning, nearly 60 years later, it was incinerated over the skies of Texas.

Possible Causes of the Disaster

The subtitle of another article appearing in the *Times*, on February 2, written by William J. Broad and James Glanz, was "Inquiry Putting an Early Focus on Heat Tiles." NASA had experienced trouble with the tile heat shields on the wings in the past, and although NASA officials discounted the effects of a piece of foam insulation shed by the solid rocket tank, they had to admit the loss of sensors in the left wing would necessitate study of that possibility. Broad and Glanz listed five other possible causes; the six possibilities, in order of decreasing likelihood:

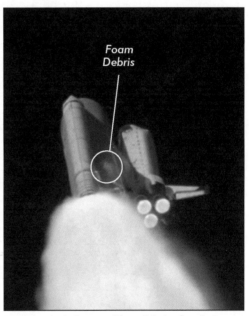

Figure 1-1. *A shower of foam debris after the impact on* Columbia's *left wing. The event was not observed in real time (CAIB Report, Vol. I, p. 34).*

1. Damage to the protective tiles on the left wing.

2. An explosion of the ship's fuels and oxidizers, which were kept under high pressure.

3. Collapse in the shuttle's structure, which was aged.

4. Faulty navigation setup for the fiery re-entry, caused perhaps by a computer problem.

5. A collision with a speeding meteoroid or piece of space debris.

6. Terrorism, perhaps by a technician at the launching site.

As we shall later see, research and analysis would proceed by an attempt to reject faulty hypotheses and seize upon the hypothesis that would explain most of the facts. Although it would take some time to prove, the answer would be that none—not one—of these original hypotheses would turn out to be the complete answer.

There was another question: Could NASA have saved the astronauts?

These questions would remain a mystery for months to come.

We can now move to Chapter Two and pick up the narrative of what was learned in the following week. ✦

Chapter Two

The Week Following: Debris, Data, and Fault Trees

But we are bound to exhaust all other hypotheses before falling back on this one.

> —*Sherlock Holmes in* The Hound of the Baskervilles, *by Sir Arthur Conan Doyle*

[Watson] "Is there any point to which you would wish to draw my attention?"
[Holmes] "To the curious incident of the dog in the nighttime."
[Watson] "The dog did nothing in the night-time."
"That was the curious incident," remarked Sherlock Holmes.

> —*"Silver Blaze" by Sir Arthur Conan Doyle*

The title of this chapter encompasses four different meanings of the word *debris*: the first is the NASA usage referring to parts of the foam insulation that fell off the external fuel tank and apparently hit the *Columbia*'s left wing; the second is the description of pieces and parts, including human ones, of the shuttle and its crew that fell on Texas and Louisiana; the third refers to the junk left in orbit around the Earth; the fourth refers to pieces of wreckage from damage that had been done to NASA-as-organization by the *Challenger* accident 17 years earlier.

This chapter is presented as the news came in on a day-by-day basis in the first seven days after the tragic accident. It will seem unstructured—for a reason. I want the reader to sense the chaotic conditions after a national tragedy. It is a time in which people reflect on the meaning of exploration, of life and death. Everyone seems to have an opinion. Reporters and official investigators jump from finding to finding and from theory to theory—old research studies are discovered and pondered. NASA positions are presented in news conferences and then reversed the next day. It is a purposive chaos, however, as people strain to find evidence bearing on the mysteries. The ethical implications of the actions of the NASA employees are always just below the surface of the information.

Monday, February 3, 2003

As the nation went back to work on Monday morning, so did the journalists. Kenneth Chang of the *New York Times,* for example, explained how re-entry works: In its final orbit around the earth *Columbia* was flying upside down in relation to Earth and backward so that it could fire its engines long enough for the craft to slow down by 175 miles an hour, just enough to let gravity take over and drag *Columbia* into the earth's atmosphere; then the pilot made it do a flip, nose over tail, so that it could descend at an angle of about 40 degrees, nose slightly up, using only its flaps and pressure of the atmosphere to slow it down. It was only an hour away from Cape Canaveral, where Gregory Williams and the others were waiting for it. As it passed over California, however, four temperature sensors in the left wing failed and temperatures in the left brake line began to rise. Mr. Dittemore, the shuttle program manager, said the problem may have been with heat tiles.

James Bennett filed a story from Qiryat Motzkin, Israel, home of the late hero, Colonel Ramon. A high school there had prepared an experiment to be carried on board the last flight of *Columbia,* an experiment to see how cobalt and calcium crystals would grow in space. The blue and white crystals represented, of course, the Israeli national colors, just as some thought the blue and white sky the day of the shuttle's last launch represented them. The students were doubly devastated by the loss of a national hero and their experiment. Dor Zafrir, a 16-year old boy in the school, was quoted as saying the voyage of the shuttle "was almost the only good thing in Israel." And: "No matter what we do, nothing comes up right," said the boy. A columnist in Israel was quoted as considering the possibility of a "curse" that had turned the Israelis into a "paranoid people." Against that possibility, he wrote, "history will remember Ilan Ramon not as the victim of a collective Jewish sin, but as another Israeli pioneer who fell."

Back in the United States, the technological failure of *Columbia* led many to search for spiritual comfort. "In a flash of fire, it became less a matter of space than of heaven." That is the first line in an article from Houston. It said that people are looking in different directions for guidance; a Buddhist is quoted to the effect that death is but a part of the circle of life. It reported memorial services held for all the astronauts in parks, synagogues, and Christian churches of several denominations, including the Olympia Brown Unitarian Universalist Church in Racine, Wisconsin, which Commander Laurel Salton Clark considered home. The Pope prayed in Rome for the astronauts.

Elissa Gootman of the *New York Times* reported that Colonel Robert D. Cabana, Director of Flight Crew Operations at NASA's Johnson Space Center near Houston, had had a lengthy conversation with the two astronauts and one cosmonaut on board the International Space Station. The two American astronauts were Captain Kenneth D. Bowersox of the Navy, commander of the station, and Dr. Donald R. Pettit, a flight engineer. The Russian cosmonaut, Nokolai M. Budarin, was a flight engineer. "They're grieving up there also, and they feel a little isolated," said Cabana. He added that they were looking forward to a visit and supplies from *Progress,* the Russian robotic spaceship which had been successfully launched the previous day.

The press seemed self-conscious about its attention to the shuttle disaster. Jim Rutenberg and David Carr write of a "hard switch" from one event to another: "It became clear that for the next few days at least, Iraq would have to fight for time and space with news about the shuttle." Diane Sawyer and Charles Gibson of ABC's *Good Morning America* left Turkey, where they were to report on the imminent American invasion of Iraq, and returned to the United States. Tim Russert of *Meet the Press* canceled interviews with political figures to devote the time to the accident. *Time* magazine published 17 pages of coverage; *Newsweek,* 20.

NASA had no independent board to investigate accidents at the time of the *Challenger* explosion in 1986; it could only investigate itself. President Reagan appointed a Presidential Commission to investigate the accident. It was called the Rogers Commission after its chairman, former Secretary of State William P. Rogers. NASA had set up an "independent" board of its own at that time, the Mishap Interagency Investigation Board. This board was convened a few hours after the disintegration of the shuttle *Columbia*. NASA announced it would be chaired by Harold W. Lehman, a retired admiral; other members included an Air Force officer, another naval officer, the director of the Federal Aviation Administration's Office of Accident Investigation, and a manager from the Department of Transportation. Almost immediately critics emerged to say the board was not truly independent because it would report to the NASA Administrator rather than to Congress and the White House and because it contained too many people connected with the government. Critics also said an investigation should delve into management, organization, and communication problems within NASA, as well as the technical causes of the disaster, questions motivated in part because of how NASA had handled such questions in 1986.

The *New York Times* printed excerpts on Monday from a news conference held Sunday at JSC in Houston. Ron D. Dittemore, the space shuttle program manager, and Robert Cabana, an astronaut and the International Space Station manager at JSC spoke and answered questions. Dittemore reviewed the facts of temperature rises as the shuttle passed over California, and then said there was much to do: "So we've got some more detective work, but that's why I say we're making progress inch by inch." Reporters raised the question of the debris hitting the wing. Dittemore said there had been a "thorough discussion" of the incident: "Our technical experts believed the debris that hit the Orbiter was inconsequential. It was not going to represent an impact to our flight control qualities or its safety." He continued, saying there was "strong participation" in the discussion from people concerned with quality, safety, flight crew members, mission ops (an abbreviation for operations), the technical and engineering disciplines, and appropriate management people. "And we concluded it did not represent a safety concern," said Dittemore. "But as we gather more evidence, certainly more of the evidence may point us in a different direction."

Reporters returned to the question of debris hitting the Orbiter at the launch, asking whether ice—it had been rainy in Florida for weeks—could have been a factor. And they pressed again and again about the debris. Mr. Dittemore: "And all these people were engaged. All of them heard the story. All of them reviewed it to their satisfaction. And the consensus, the *unanimous consensus*, was that it was as I represented it to you earlier, it was not a significant event" (emphasis added). *(Could it have been a false consensus?)*

One of the more important articles of the day was by John Schwartz, a piece about how the heat shield, including the protective tiles on the shuttle, had been a major concern since the start. The start was the first trip *Columbia* took, not into space but strapped, piggyback style, on the top of a Boeing 747 on a flight from California to Florida. It lost 40 percent of its 5,000 most critical tiles during the trip. It lost a dozen tiles or so on its maiden flight into orbit in 1981. The tiles are vitally important to the shuttle by providing, with other materials, a heat shield allowing the craft to survive the deep freeze of space and the high temperatures during re-entry.

The skin of the shuttle is made of aluminum; this skin is covered with a layer of flame-retardant nylon felt; the tiles, about six inches square and one to five inches thick, are applied to the felt much like a jig-saw puzzle. The nose and leading edge of the wings experience the greatest amount of heat, so they are covered with a light-gray material called reinforced carbon-carbon, or RCC. It is much too heavy to put on the entire vehicle, so the tiles, as light as balsa wood, are used on other parts of the machine less vulnerable to the stress of heat changes. According to Schwartz, a study released in 2000 said NASA hadn't done enough to guarantee safety for the shuttle, suggesting that there might be the possibility of corrosion under the tiles. Warren Leary reported that *Columbia* was not only the oldest shuttle, she had a reputation for quirkiness on the launch pad and was nicknamed the "hangar queen" because of time in the shop for repairs.

The topic of the foam debris wouldn't go away. James Glanz and David Barstow contributed an article about loose foam that began with this two-sentence paragraph: "When insulating foam is applied to the space shuttle's 15-story high external fuel tank at the manufacturer's plant in New Orleans, it goes on like shaving cream, soft and gooey. But after it cures the foam turns hard as brick." They also interviewed some scientists not associated with NASA who were critical of the space agency's decisions. An expert on the foam said it "was hard, like an old-fashioned brick." Other scientists were critical of NASA's decision to rule out damage from the "scant evidence" they had. NASA's video cameras were relatively ineffective during the launch, the most important one being out of focus, and yet they didn't request other agencies to use their equipment to look at the shuttle after the launch. The Defense Department, for example, had two telescopes that could have taken pictures of tiles had they been asked to do so. And there were also the spy satellites, about which people were reluctant to talk but which could have been used to photograph the shuttle.

William J. Broad and Carl Hulse dug into NASA's history to write an article with this subtitle: "NASA Dismissed Advisers Who Warned About Safety." The advisers were members of NASA's Aerospace Safety Advisory Panel (ASAP), a group of experts from industry and academia. Five members of the group were removed one year before the *Columbia* accident after they warned that cuts in NASA's budget could have serious implications for the safety of the shuttles. A sixth member quit in anger after the other five had been fired. One of the concerns the panel had was cuts in the NASA workforce. Some felt that technical competence within the agency was becoming thinner and thinner. Sonja Alexander, a spokeswoman for NASA, said of the removal: "It had nothing to do with shooting the messenger."

"Shooting the messenger" is an updated expression of an ancient technique for dealing with bad news. The Greeks are said to have killed messengers who brought bad news from remote military battles. They did it not so much for punishment as to eliminate the possibility that others, including enemies, might extract the bad news from the messenger. It is not likely that they would "shoot" the messenger, given the absence of guns in those days, but spokeswoman Alexander was thinking in real time. "Killing the messenger," at any rate, may help the leaders of an organization to avoid embarrassing moments, but it is bad for the organization, its members, owners, and clients. It is also a sin of organizational communication, a deadly transgression.

Reflecting on the news of the day, I remembered a famous essay on language written by an anthropologist and linguist, Benjamin Lee Whorf. Whorf studied chemical engineering at the Massachusetts Institute of Technology and became an investigator for an insurance company. After that he began the study of languages, including those spoken by Native Americans, in order to discover linguistic principles. The essay I recalled, "The Relation of Habitual Thought and Behavior," drew on his experiences as an insurance investigator to explain what he'd learned studying languages. It has been called the Whorf Hypothesis, sometimes the Sapir-Whorf Hypothesis to acknowledge one of Whorf's teachers, Edward Sapir.

The Hypothesis was stated this way by Sapir: "The fact of the matter is that the 'real world' is to a large extent unconsciously built up on the language habits of the group" (as quoted in Whorf 1956, 134). The language of the group structures reality. In a section of his essay with the heading, "The Name of the Situation as Affecting Behavior," Whorf drew on his experience as a fire investigator to illustrate the idea. His investigations were directed toward purely physical conditions such as defective wiring. In due course, however, he discovered "that not only a physical situation *qua* physics, but the meaning of that situation to people, was sometimes a factor, through the behavior of the people, in the start of the fire" (Whorf 1956, 135). Another way of stating the principle is that words used to characterize a situation or thing can influence the meaning they have for people in a *habitual* way.

Whorf proceeded to illustrate this claim with two examples of situations in which names affected behavior in a habitual way. The first had to do with gasoline. Around objects with the name of "gasoline drums" workers were quite careful. Around "empty gasoline drums" workers were extremely careless with matches and cigarette stubs, starting fires that became conflagrations. "Empty" gasoline drums, of course, can be "full" of dangerous vapor. Drop a match into an "empty" can like that and you may not be around to understand the danger.

The second example comes from an investigation in a wood distillation plant. The metal stills were heated to extremely high temperatures. They were insulated with "spun limestone." Workers made no attempt to protect the *insulation* from excessive heat or contact with flames. Imagine how surprised the workers were when the lime*stone* turned out to be combustible. Everyone knows "stone" doesn't burn—yet it did.

I began to cultivate a hypothesis, an idea that the relative inattention to the problem of the loose foam might have had a semanto-genic origin; that is, the common meaning of *foam* might have discouraged people from taking it seriously. Everyone knows that foam is light, particularly the foam insulation and packing material around us so much of the time. Everyone knows that you could throw a chunk of that at a shuttle without any effect at all. People, even engineers and scientists, might develop a habitual response to it. *But what if the chunk was as hard as a brick?* If Whorf found many instances of this response in the fires he investigated, it is clear that language—names of objects and situations—could give people a "preunderstanding" of a situation or a material that could lead to a misunderstanding. Inappropriate behaviors would then follow from misunderstandings. Whorf's ideas are considered later in our analysis of the accident.

A *New York Times* editorial considered the investigation of the accident, recalling how NASA had infamously tried to cover up its lapses in the early stages of the *Challenger* investigation. It charged that the NASA board in charge of the *Columbia* investigation had neither the status nor the independence to have much credibility. "An independent presidential commission with distinguished members from the private sector investigated the *Challenger* accident. President Bush should appoint a similar panel to investigate this one while Congress pursues its own inquiries." In an article by Matthew L. Wald, a different perspective on the Rogers Commission emerges. Wald quotes John C. Macidull, a former staff member of the Rogers Commission, about how much could be learned from the wreckage of *Columbia*, and later identified Macidull as the author of a book published the previous December with the title, *Challenger's Shadow: Did Industry and Management Kill Seven Astronauts?*

A delightful Op-Ed piece by Timothy Ferris of San Francisco told about watching the seven astronauts the previous Thursday on the Internet, one of their live broadcasts from space. "They demonstrated how they ate their favorite foods in the weightless environment (carefully, to prevent crumbs floating around), wryly displayed the frozen blood samples they were bringing back for laboratory analysis on the ground, and cavorted in weightlessness as delightedly as otters on ice." Ferris was drawn to Mission Specialist Laurel Clark of Racine, Wisconsin, whose gentle and effervescent demeanor seemed inconsistent with her occupations as Navy Seals diver and Navy Flight Surgeon. She happily reported that a moth she was observing had started to pump its wings.

" 'Life continues in lots of places,' she reflected, 'and life is a magical thing.' "

Tuesday, February 4, 2003

By the next day the *debris at liftoff* had become a "leading suspect" in the mystery of the lost shuttle in the subtitle of an article by John M. Broder. Mr. Dittemore said the external tank, and its insulating foam, was suspected of being the "root cause" of the disaster. Mr. Dittemore is quoted as saying that some people in the organization had expressed reservations the day after the launch, saying that the debris strike was a potentially serious problem. NASA encouraged such dissent, he said, but as shuttle program manager, "he had not heard them." *(This is another example of a potential communication transgression in which dissent is said to be encouraged yet isn't heard.)*

Another article turned up a 1997 report that had warned of foam debris as a cause of damage to the ceramic tiles protecting the shuttle from re-entry temperatures. James Glanz and Edward Wong reported that a NASA engineer at Cape Canaveral had said in a report dated December 23, 1997, that more than 300 tiles on the *Columbia* had been damaged on a recent flight. The engineer concluded that the damage was "not normal." A videotape of the most recent launch, January 16, 2003, showed a basket-sized chunk of debris breaking away, striking the left wing, and exploding in a cloud of dust. The chunk seemed to be about 40 inches by 40 inches. Some experts were quoted as saying roughness of the tiles could create a "turbulence" during the re-entry and thus a dangerous heating of the Orbiter's aluminum skin. Mr. Dittemore said, however, there had to be a "missing link" they did not understand at this time.

Could NASA have helped rescue the seven astronauts if they had known there was damage to the left wing? Reporter Kenneth Chang raised that question on February 4 in his article, the subtitle of which gives the answer: "*Columbia* Was Beyond Any Help, Officials Say." There is a rescue procedure available when NASA officials know something is wrong; they can drop the shuttle's solid rocket boosters and the external fuel tank and glide back to earth. But no one knew the foam had hit the wing until the next day when engineers were making a frame-by-frame study of a videotape of the launch.

Four ideas about rescuing the astronauts have been mentioned, wrote Chang, since the accident last Saturday, three of which Gregory Williams and I enumerated in our interview. First, assuming there were damaged tiles, could the crew have taken a space walk and fixed them? No, they had no tools. Moreover, NASA rejected that idea when they decided the astronauts could have done more damage than good by climbing around on the bottom of the shuttle.

Second, could the shuttle have docked at the International Space Station once the damage was discovered? No, after dropping its external tank, *Columbia* wouldn't have had sufficient fuel to climb to the orbit of the ISS at a higher altitude. "We had nowhere near the fuel needed to get there," said a NASA spokesman at KSC.

Third, could another shuttle scheduled to carry supplies and a new crew to the space station on March 1 be sent to rendezvous with *Columbia* and rescue the crew? No, the *Atlantis* was still in its hangar and in need of three more weeks of preparation for launching, while *Columbia* had enough oxygen and supplies for only five more days. Rushing the *Atlantis* would have compromised NASA's regulations, endangering its crew as well.

Fourth, could the angle of re-entry have been altered in some way so as to let the right wing bear the brunt of the heat, sparing the damaged left one? No, said Mr. Dittemore on February 4: "I'm not aware of any other scenarios, any other techniques, that would have allowed me to favor one wing over the other." Chang also quoted Gene Kranz, the flight director who managed the rescue of *Apollo 13*, who said there were no options. Gregory Williams and I must have been wrong in our rescue scenarios. Or were we? Chang did paraphrase Dittemore as saying: "Even if that had been possible"—favoring one wing during re-entry—"it would probably have damaged the shuttle beyond repair and made it impossible to land, requiring the crew to parachute out at a high speed and at high altitude. He said there was no way managers could have gotten information about the damaged tiles that would have warranted so drastic a move."

(Parachutes?)

The *New York Times* ran part of NASA's Twelfth Day Report of January 28, 12 days after launch; I reproduce these lines from it: "The impact analysis indicates the potential for large damage area to the tile and reinforced carbon carbon or RCC. Damage to the RCC should be limited to coating only and have no mission impact."

During a press conference printed Tuesday a reporter put this question to General Michael C. Kostelnik, deputy associate administrator for the space shuttle and International Space Station programs:

> Q: General, there's a memo that surfaced this morning that goes to your statements about concern about the astronauts and the people. It suggests that somebody in your operation knew about extensive tile damage, wrote a memo two days before the accident. And my question is can you give us the chronology of how that memo was handled and how high in the organization did it get?

General Kostelnik: That's actually the first I have heard of that memo. . . . This new report is a new one to me. If we can get the details on it we'll run it to ground and I will have the program this afternoon present what we know.

The afternoon session of the news conference was handled again by Mr. Dittemore. He spent most of the time reviewing the foam incident, analyses of it, and the decision, despite some observed damage to tiles, that it was not a risk to the safety of the flight. And then he was asked a personal question: Was it therapeutic to come back to work on Monday after such a "wrenching weekend"? Mr. Dittemore:

The hardest thing that I've had to do over the past two days was drive home in my car Saturday afternoon alone with my thoughts. I've talked to several others that had the same experience. As long as we are together, as long as we are trying to solve the problem, we can stay focused, we can keep our energy directed in the right direction. *~team mentality*

Later he was asked, "After looking at all the reports from engineers that concluded there wouldn't be a problem who was the final person who signed off on it?"

Mr. Dittemore: The person with the authority to judge whether it represents a risk or not ultimately is the chairman of the mission management team. However, I'm the accountable individual. Anything that happens at the mission management team I am made aware of. . . . So ultimately it's my decision. And it doesn't matter what anybody else thinks. If I believe it's not a safe thing to do, then we won't proceed. In this case, I did not chair the mission management team. It was delegated to the manager of the program, manager for program integration, who was also serving as the launch integration manager for launch. But I was kept informed and knowledgeable at all times.

(He was accepting responsibility for the tragedy.) I was not unimpressed. Dittemore's words, whether he was to blame or not, marked a refreshing departure from the precedent set when NASA administrators distanced themselves from the decision to launch the doomed *Challenger* and to this day deny responsibility and evade any kind of sanctions.

Reporters were quick to remark on how differently NASA was handling this case in comparison to the previous shuttle disaster.

Jim Rutenberg recalled how NASA answered no questions during the two weeks following the *Challenger* explosion, not even a question about the ground temperature at the time of the launch. Even after a flame had been detected coming from the booster, the NASA Public Affairs people could not use the word *flame*; they had to refer to it as an *anomalous plume*. That is a form of denial.

John Noble Wilford, a long-time science writer for the *New York Times*—he covered the Apollo Program and wrote a book about it— explained the rationale for both the shuttle program and the International Space Station. Wilford quoted Alex Roland, a historian at Duke University, as saying, "There's not much good reason for the shuttle except to go to the space station, and not much reason for the station except to give the shuttle a place to go." Roland did admit he was being deliberately cynical, but others share his opinion.

Some see, however, that good has come by having the former Cold War enemies—the United States and Russia—cooperate in building, along with several other countries, the International Space Station. It gave the former Soviet space team something to do (as it did the shuttle) other than sell their knowledge to other countries. The Russians have two kinds of spacecraft that can dock at the station, one that carries cosmonauts—the *Soyuz*—and one that is robotic—the *Progress*. But the Russians began to run out of money, forcing the United States to spend more than was expected on building and supplying the station. In fact, the new Administrator of NASA, Sean O'Keefe, in his job for only a year, had spent much of the previous 12 months trying to get a handle on the cost overruns created by the space station. O'Keefe is reported to be a political protege of Vice President Dick Cheney, and the Bush administration proposed an increase in NASA's budget from $3.2 billion to $3.97 billion this year, even before the *Columbia* disaster.

Critics of the space program argued once again that NASA should not try to fly humans into space; rather, it should explore the vast unknown with robotic space ships. One historian, John Staudenmaier, is paraphrased in an article by Amy Harmon as saying that the combination of terrorist attacks on September 11, 2001, the collapse of the Internet bubble, and the shuttle disintegration have undercut the technological bravado that has driven U.S. exploration: "There is this feeling of vulnerability right in the center of where Americans feel most confident, our sophisticated technological systems. For Americans, it's almost like being homeless."

In defense of manned flight, an essay from the *New York Times* Science Desk by Dennis Overbye appeared on February 4. He had covered shuttle flights back in the '80s and had been amazed at how much noise and fire and fury it took to get a few human beings out of the grip of Earth; he wondered whether it would be possible that so much "violence" could ever become routine. While *Columbia* was breaking up, Overbye was 20 feet underwater in the Caribbean, trying to learn how to take his diving mask off and put it back on under water. It's a skill our descendants may need to learn, he wrote, because a new book on his desk back in New York argues that all life on Earth will have to return to the oceans in a half-billion years. The book, *The Life and Death of Planet Earth,* by Donald Brownlee and Peter Ward, predicts that it will be too hot to live on land when the sun's luminosity increases. And then in about two billion years the oceans will boil, sterilizing the earth. It may not take that long if an asteroid or planet hits the Earth. Either way, there will be no one who remembers life on Earth, or any of us for that matter, unless our descendants leave the Earth and find other places to live. If they do, they might think of Earth forever after as Eden or perhaps even as Heaven.

There was some good news about the last mission of *Columbia*, a lasting legacy to the seven astronauts. Many experiments and specimens were lost in the break-up, but the astronauts, according to Warren E. Leary of the *New York Times*, "had transmitted a wealth of information from a number of experiments back to earth during the 16-day research mission." He quotes a professor at the University of Michigan, Gerard M. Faeth, as saying he and his colleagues got high-quality data before the disaster. They planned to acknowledge the cooperation of the astronauts in future publications. Speaking of the seven astronauts, Faeth said: "They were our colleagues, part of our laboratories. The price paid for this data was very high—seven lives lost—and we are grateful. These were people we liked."

Other scientists would later disagree, would argue that such "shuttle research" is either of only marginal scientific value or could be conducted just as well on unmanned flights. The astronomers would make a different point, arguing that they need manned flight to keep NASA's orbiting telescopes in orbit and in working order. The issues on this question are complex.

Wednesday, February 5, 2003

Fewer stories were printed each day because reporters were catching up, to some extent, with the need for supplying background information to their readers. How many times can you write informative articles about the tiles and the RCC? Carl Hulse did report on February 5 that Democrats and Republicans in Congress were eager to expand the 10-member Mishap Interagency Investigation Board because it was dominated by government and military officials. There was a veiled threat by Representative Bart Gordon of Tennessee, senior Democrat on the House Science Committee:

> If there isn't more private sector expertise brought in, I think you will find a strong call for a so-called blue ribbon committee. I don't think we need three different commissions doing the same thing. It would be preferable to make this outside board truly outside and credible.

NASA had come up with new expressions about the area in which pieces and parts of the shuttle had landed: the "debris field," or the "field of debris." NASA suspected a wider debris field, one that included California as well as Texas and Louisiana. Matthew Wald and Andrew C. Revkin reported that in East Texas people were lining up shoulder-to-shoulder to comb the pine woods for debris. The search was in an early stage, "bag and tag," so the debris could be taken to Barksdale Air Force Base near Shreveport, Louisiana. The next step would be to attempt to reconstruct the shuttle itself; any pieces from the left wing would be treated as extremely important.

Another new word popped up in articles: *bipod*. A bipod is an attachment point that connects the external fuel tank to the shuttle's nose (see Figure 2-1). It is in the shape of the letter V, and if you turn it upside down it appears to have a crotch and two legs. Perhaps that's what inspired some engineer unknown to me to call it a bipod. *(Why not biped?)*

Reporters Broad and Sanger found scientists who had done research for NASA about the foam insulation and the bipod. While NASA was saying they doubted the foam could have done damage, the reporters found that the space agency had paid for a study in 1990 by researchers at Stanford and Carnegie Mellon universities. The study showed that the wheel well areas were vulnerable

Figure 2-1. Computational Fluid Dynamics was used to understand the complex flow fields and pressure coefficients around bipod strut. The flight conditions shown here approximate those present when the left bipod foam ramp was lost from External Tank 93 at Mach 2.46 at a 2.08-degree angle of attack (CAIB Report, Vol. I, p. 52).

to being hit by debris and always experienced high temperatures during re-entry. The loss of even one tile could have a "zipper effect," stripping away other tiles. The loss of tiles could allow plasma, hot gases, to penetrate the aluminum skin of the shuttle.

The reporters quoted the researchers as saying that NASA had recently called to ask for copies of the study it had received and paid for 13 years earlier. The study found that ice as well as hardened foam could cause damage to the tile. The area where the bipod pierced the foam was thought to be dangerous; the foam might pull away more easily from the fuel tank at that point. Again the foam was described as shaving cream that "turns as hard as a brick." General Kostelnik, who had joined NASA the year before, said he was not aware of a particular safety risk associated with the wheel wells.

A rock star became part of the story when the Russian Space Program had to shut down its tourist program because of the emergency caused by the shuttle disaster. They had made some money, according to Michael Wines, by flying rich tourists into space and back, but now their *Soyuz* craft would be needed to service the International Space Station. American space officials had frowned on the program until a millionaire from South Africa and a California financier took separate trips on *Soyuz* capsules. They got even more interested when a singer named Lance Bass of 'NSYNC announced a year ago that he would fly on the *Soyuz* to the space station; the singer's plans fell through last fall when he and the Russians were unable to agree on the price of a ticket.

Articles abounded during the week about the economic impact of the second shuttle disaster. Florida, for example, is highly dependent on space spending, 92 percent of which goes to private contractors. Much money was lost to the state during the period of nearly three years when the shuttles were grounded in order to fix the O-rings after the *Challenger* disaster. New Orleans, home of the

Michoud Assembly Facility (where the external fuel tanks are made) suffers from unemployment whenever there is a slow-down in shuttle flights. Houston's economy is affected by the ups and downs of space flight because of the presence of JSC and many contractors. NASA's plans to develop a reusable space plane, an orbital space plane, to take astronauts and supplies to the space station could mean a boost for the aerospace companies, even though they now faced competition from Russia and three newcomers, Japan, China, and India.

Thousands in Houston finally had a chance to mourn collectively at a service for the seven fallen astronauts. The family theme was expressed again and could be inferred from NASA employees who attended dressed in black and wearing sunglasses to cover red, teary eyes. According to Kate Zernike and Nick Madigan, this was a funeral for a family of thousands. Captain Kent V. Rominger, the chief of the astronaut corps, read the huge crowd a verse from Joshua that Rick Husband had read to the crew and their families the night before the launch: " 'Have I not commanded you, be strong and courageous,' he read. 'Do not be terrified, do not be discouraged, for the Lord your God will be with you wherever you go.' " Rominger's voice broke during his conclusion: "Rick, Willie, Mike, K.C., Laurel, Dave, and Ilan, I know you're listening. Please know you are in our hearts, and we will always smile when we think of you." President Bush also spoke, mentioning each of the seven fallen astronauts by name and speaking about the grim reality that some explorers don't return.

Thursday, February 6, 2003

The plot thickened in an article written by James Glanz. The subtitle: "MYSTERY DEEPENS." The first sentence: "NASA investigators said yesterday the disintegration of the space shuttle *Columbia* had turned into a scientific mystery." The article said that the facts available to investigators didn't seem to "jibe." There was no explanation for what was known: the demise of the shuttle. Computer simulations had failed to show how a piece of foam insulation could cause the catastrophe. Data from sensors didn't record the soaring temperatures required to melt the aluminum skin of *Columbia*. Video images showed pieces of the shuttle dropping off long before it had troubles in flight. The facts didn't add up.

A welter of theories had been generated: a hit on the wing by a tiny meteorite during descent, knocking out sensors and preventing the craft from withstanding the 2,000-degree temperatures; irregularities or minor damage on the tiles, producing turbulence in the flow of air over the wings; the foam-tile theory in which gases entering the wheel well set off explosive devices called pyrotechnics (devices there to blow off the landing gear doors in case they don't open for a landing).

Mr. Dittemore said in a news conference at JSC that NASA was backing away from what had been their leading theory—that foam breaking away from the external fuel tank had damaged the ceramic tiles on the wing. Their calculations told them that at 1,000 miles per hour, twice what they believe to be the actual rate, a piece of foam would do only inconsequential damage to the tiles. That was diametrically opposed to what Dittemore had said 48 hours earlier when he expressed the assumption that shedding from the external tank was the "root cause" of the damage that doomed *Columbia*.

Setting aside the question of the tile, another riddle had NASA stumped. It was not evidence but lack of evidence that was perplexing: The sensors recorded a rise in temperature of only 40 to 60 degrees; the aluminum skin would have been exposed to temperatures of nearly 3,000 degrees, and aluminum melts at 1,200 degrees. "The absence of any observed high temperatures has taken on the role not unlike the famous dog that did not bark in the Sherlock Holmes story—an absence that has to be explained before the mystery of the *Columbia*'s demise can be explained," wrote reporter Glanz, alluding to a story by Sir Arthur Conan Doyle, "Silver Blaze." In that short mystery, Sherlock Holmes is puzzled by newspaper accounts that a famous racehorse, Silver Blaze, so named because of a white patch on his forehead, had been stolen before an important derby. Holmes solves the mystery by realizing what did not happen: The fact that the dog guarding the stable didn't bark when the horse was stolen meant that the horse thief was well known to the dog.

Sherlock Holmes is still celebrated for his powers of deduction. Engineers use a similar but more formalized procedure, reverse engineering by means of fault tree analysis. Mr. Dittemore explained his method in a press conference published on the same day the Sherlock Holmes analogy appeared in the *Times*. He also seemed to imply they were trying to exhaust all other hypotheses before dealing with the question of the foam. After explaining why he and the other engineers at NASA couldn't accept the theory that foam shedding from the tank could represent an issue of flight safety, Mr. Dittemore said:

> So we're looking somewhere else. Was there another event that escaped detection? As I mentioned before, we're trying to find the missing link. And as you focus your attention on the debris, we're focusing our attention on what we didn't see. We believe there is something else. And that's why we're doing a fault tree analysis. . . . And remember what a fault tree analysis is. You start at the top by saying you lost the vehicle. And then underneath that, all of the reasons you might lose the left wing.

As Sherlock Holmes did so brilliantly in his head and in conversations with his sidekick, Dr. Watson, the NASA engineers were doing graphically, drawing a tree with many branches, many possible causes, trying to eliminate false leads and come to agree on the cause (see Figure 2-2).

Mr. Dittemore also said in the press conference that he had talked to the ice experts and ruled out that water could have been absorbed by the foam: "There was a question also about whether or not foam absorbs moisture. It does not. It's very resilient to any rain or moisture. And logically you would believe that to be true. If it weren't, it would be absorbing rain and moisture and it would become much heavier." He did acknowledge that ice could form on the fuel tank, but they had strict criteria about ice; experts were sent out to inspect the tank after it was filled. So the problem couldn't be ice. He also said that NASA engineers had expressed "reservations" about the situation but said that their concerns were about "process," not about "conclusions."

Reporters picked up on the remarks about fault tree analysis; Broad and Revkin explained it to their readers in the subtitle of their article: "THE PUZZLE: Engineers List All the Ideas, Striking Them One by One." Fault-tree analysis is like drawing a map, they explained: start with what you know, a shuttle lost as it returned to Earth, and then list every possible scenario of what could have caused the loss, and then see which plots have more and less support from the engineering data, a process of elimination. A large branch of the tree is, of course, the left wing and the wheel well where sensors registered rises in the temperature. The increased temperatures were not enough to destroy the craft but were high enough to become what NASA likes to call "anomalies."

A similar analysis was being made to explain the craft's last few movements, sideways movements in which *Columbia* was "struggling to compensate for a mysterious force that was dragging at the left wing." NASA continued to try to prune branches off its fault tree.

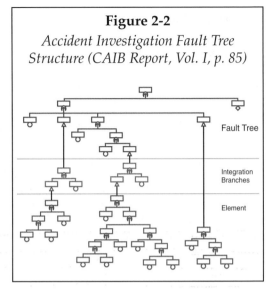

Figure 2-2

Accident Investigation Fault Tree Structure (CAIB Report, Vol. I, p. 85)

Fault Tree

Integration Branches

Element

A short article by reporter Wong announced in its subtitle the arboreal theme: "Object, Caught in a Tree, Prompts Thoughts of God." But this turned out not to be a fault tree. Nancy Youngs was sitting in her backyard swing with Bible in hand when she saw a white fabric about 30 feet up in the branches of an oak tree. She lives not far from the Michoud Assembly Facility where the external tank for the shuttle is manufactured. She wondered if it could be debris from the *Columbia* breakup. Other pieces of debris had been reported in the area.

"That would be God giving someone a wake-up call," Mrs. Youngs said.

Congress announced hearings for the following week in which NASA Administrator Sean O'Keefe could testify before committees of both houses rather than meet twice with them separately. We learned also that families of the seven astronauts who perished in *Challenger* received millions of dollars from the government and Morton-Thiokol, the contractor who made the solid rocket boosters. There were legal reasons, however, why that might not be a precedent for the families of the seven lost in *Columbia*. On another legal matter, Federal agents arrested two people in East Texas and charged them with shuttle debris theft. One person took a piece of thermal coating from a soccer field and another, posing as a NASA official, took a circuit board from the person who found it. The government wanted to make an example and warned others that they would have two days of grace in which to turn over debris without penalty or charges—a kind of amnesty. Heavy rains in the debris field made it tough going for police officers on horses and others walking through muddy fields. Remains of astronauts were taken to the military mortuary at Dover Air Force Base in Delaware.

Friday, February 7, 2003

An overview article (by Broder and Oppel) contained a couple of surprises. NASA, under pressure from criticism by Congress and other sources, announced it would relinquish authority over the board reviewing the shuttle accident and agreed that new members would be named to the panel. David Goldston, the House Science Committee Chief of Staff, said Republican committee members wanted broader voices on the panel, even though they were "agnostic" about who it should report to.

In a press conference reported on Friday the shuttle program manager, Ron D. Dittemore, who had been handling briefings with a "certain grace," appeared "weary and befuddled" as he met with reporters in an auditorium at JSC in Houston. "He admitted he and his team were emotionally and physically drained, as well as frustrated by the poor quality of the evidence they were studying." They couldn't get a good view of the debris strike because of an out-of-focus camera. He retreated from the assertion he had made on Wednesday that it was unlikely a piece of foam could have doomed the spacecraft. No potential cause had been ruled out; all possible theories would be pursued. The fault tree, thus, still had all its branches.

In two separate press conferences the name of the newly independent investigating panel was used, in one held by the panel and another by NASA Administrator Sean O'Keefe. In the first there was a self-reference by Admiral Lehman to the Columbia Accident Investigation Board. In the second, O'Keefe used the same designation—not the Interagency Mishap Investigation Board—as part of an attempt to quiet critics in Congress about the independence and makeup of the panel. Lehman said the Board was demonstrating its independence by adding new members. The panel would soon be known by its acronym: CAIB.

Wald and Glanz provided their readers with an interesting definition of an airplane that could apply to the shuttle as well: "a collection of spare parts flying in tight formation." Perhaps it applied even more accurately to the shuttles, when one remembers that NASA frequently cannibalized parts of one machine on the ground to put into another to make it ready to fly. The main point of the article was to review the methods of investigating airplane disasters for clues as to how the analysis of the debris might be handled.

The hunt was still on in the field of debris. Jeffrey Gettleman reported from Hemphill, Texas, that the shuttle's nose cone had been found in a forest, along with astronaut remains. Over 500 military personnel were involved in the search in this area; citizens said they saw objects as big as automobiles plunge into the local reservoir. Helicopters saw shiny objects in the bottom of the body of water. Then came the rains, turning the clear water into "chocolatey darkness." The weather grounded the helicopters, beached the divers, shortened the searchers' shifts, and made people "cold, wet and miserable." Seventeen people did come in with fragments of the *Columbia* under the amnesty conditions.

A profound argument—a difference in the philosophy of science—emerged on February 7. An important article with the subtitle "NASA Seeks Answers From Simulators, Amid Some Doubting," by Andrew C. Revkin, presented competing philosophical positions. One group in NASA was proceeding to understand the cause of the accident by means of computer simulations and mathematical modeling. Others doubted that NASA, after years of strained budgets, had the right stuff to get the job done. Modeling was going to be critical to the solution of the problem, said one side. How did plasma, or hot gases, burn up a wing without the sensors showing a catastrophic increase in temperature? Supporters of the space agency said their attempts to develop codes, numeric codes, to mimic the real world were among the best efforts ever at modeling.

On the other side, Revkin interviewed a skeptical whistleblower of the past, Allan J. McDonald, who had retired 18 months earlier from Thiokol, the company that made the *Challenger* solid boosters with the infamous O-rings. He and Roger Boisjoly, a Thiokol engineer, had argued against launching the *Challenger* in cold temperatures—ice was present at the launch pad—because they had no data to indicate it was safe. He was overruled, the shuttle exploded, and he worked on the redesign of the O-rings. His position might well be called the Empirical approach, in contrast to NASA's Modeling approach. McDonald told Revkin that NASA couldn't rule out the foam theory because, although they had modeled the possibility, the agency "did not do physical laboratory tests to prove that foam at high speeds could harm tiles." Testing, he said, would "give you a far better idea of the possible damage and let you make real judgments." NASA could not, it was reported, do the tests for lack of funds. In Chapter Four we shall see how testing was an important part of what is called the Original Technical Culture (OTC) of NASA and in a later chapter how a test helped solve the *Columbia* mystery.

On the modeling side of the argument, a senior engineer at NASA's Langley Research Center in Virginia, talking with the understanding of anonymity, "emphasized that the ultimate test of the models had been the actual flights, which have shown that the models work." This is an allusion to the 111 successful flights of the shuttle, and the assumption that the data from those flights provided tests of the hardware.

McDonald talked to another *Times* reporter, D. E. Sanger, for a February 7 story about analogies between the two lost shuttles— and about the fault-tree analysis. In the case of the *Challenger*, "NASA and the company that manufactured the booster rocket, Morton-Thiokol, were littered with dissenters who warned that ignoring repeated evidence of the O-ring failure was tantamount to inviting disaster." There is a similarity between the O-rings and the tiles. Neither one is redundant; that is, neither has a backup system in case it fails. And so McDonald said:

> I would say you have to prove first that it's not the tiles, and then move on to other theories. There's a great temptation in these cases to search for some other, strange explanation—in the first days after *Challenger* I sat through meetings where people listed 30 things it could be, and 90 percent of them were ludicrous. I wanted to find reasons that it wasn't my booster rocket, he said, but, in the end, he couldn't.

In the history of philosophy there is a fascinating argument between the Idealists and the Empiricists. René Descartes (1596–1650), for example, came in from the cold to warm himself. He said "*Cogito, ergo sum,*" Latin for "I think, therefore I am." This self-evident realization led him to conclude that the only thing we can know to be true is a clear and distinct idea. Thus, we got Idealism. In opposition are the Empiricists, most notably John Locke (1632–1704), who argued that knowledge is gained by sense experience or observation. I put the Modelers in the Idealist camp and those who wanted to test the foam in the Empiricist camp. *(Which group is right?)*

Eileen Collins, 46, a veteran of three shuttle missions, had carried her 2-year-old son downstairs on the previous Saturday morning, reported Jim Yardley, expecting to see a routine landing when she turned on NASA's television channel. She watched in agony and got some toys for her son to distract him, glad he couldn't understand. "Shuttle, shuttle," he was saying. She made some phone calls and later drove to JSC. As a former chief of the Astronaut Safety Branch and an Air Force colonel, she wanted to volunteer for the team forming to investigate the accident. She got a polite refusal because unless there were a change, she was scheduled be the commander of the next shuttle flight, the mission of *Atlantis* to retrieve the three men from the International Space Station and deliver a new crew.

She has two firsts to her credit, the first woman to pilot a shuttle and the first to command a shuttle. Her role in the program gives her validation and satisfaction. She believes in space exploration by humans: "I believe we need to go into earth's orbit. We need to go back to the moon and build a space station. We need to go to Mars and build a space station." She envisions a distant future in which space travel might be necessary for the continuation of human life. In the short term, "My job is to make sure my crew is ready to fly."

Saturday, February 8, 2003

Reporter Revkin maintained the mystery metaphor:

> A week after the shuttle *Columbia* disintegrated, investigators and aerospace experts are developing an explanation for a mystery that has troubled them from the start: how onboard sensors could show isolated signs of slow warming but no sign of the source of the heat—much less the kiln-hot temperatures that would be expected with any malfunction serious enough to threaten the craft.

As part of the search for "clues" two hunches were advanced. Mr. Dittemore of NASA had showed the previous evening on a series of charts that heat sensors in and around the left landing-gear well indicated the temperature going up seven minutes before the disintegration; at the same time other sensors in the left wing and well stopped sending data. Mr. Dittemore said it was too early to know whether it was an effect or a cause, but the sensors that failed were connected by the same bundle of wires. The hunch is that some unknown force cut all the connections at once.

Another hunch came from engineers at the Massachusetts Institute of Technology: They found that a 30- to 50-degree warming, "like that recorded over five minutes in some brake-line sensors, could be generated in one part of an aluminum aircraft frame by a heat source of several thousand degrees about a yard away." Still, said Mr. Dittemore, it will be hard to cut branches off the fault tree for a long time to come.

We learned more about the bipod and the insulating foam from an independent scientist (Russell Seitz), who testified before the investigative board and gave a copy of his testimony in the form of a letter to the *New York Times,* the contents of which in turn are described by William J. Broad. Seitz told the board that NASA might have overlooked the changes in materials such as the foam when their temperatures are lowered by several hundred degrees. The liquid hydrogen in the fuel tank is cooled to minus 423 degrees Fahrenheit; the bipod abuts the tank where the hydrogen is stored. That super cold temperature could drastically increase the stiffness and strength of the foam, increasing its "ballistic potential"—or its power to damage the shuttle. NASA estimated the chunk to be 20 inches by 16 inches by 6 inches thick, which would make it the largest piece ever to strike a shuttle.

The investigative panel headed home for more clothing, having spent the week living out of small suitcases put together hastily when the shuttle failed. The board members had spent much of the week listening to NASA experts brief them about the shuttle and the possible causes of the accident. Sean O'Keefe, NASA Administrator, said on Thursday that the space agency had transferred control of the investigation to the board. He added that other members would be added: the first was Roger E. Tetrault, of McDermott International. A Representative from Tennessee, Democrat Bart Gordon, and Representative Sherwood Boehlert, a New York Republican, expressed concerns about the composition of the board.

A crowd of 8,000 NASA employees and friends gathered at the shuttle landing facility at KSC, honoring the fallen astronauts. Sean O'Keefe spoke, as did Florida Governor Jeb Bush. Many cried during the service, which ended with a "missing man" formation flown by four NASA jets.

James Glanz reported some cryptic clues, including a photograph taken of *Columbia* just before its crack-up. The photograph, taken by an Air Force telescope in New Mexico, appeared to show a slight "jaggedness" in the left wing of the shuttle. Ron Dittemore said the poor quality of the photograph wouldn't allow NASA to draw any conclusions. Even if the photograph indicated damage, said Dittemore, you wouldn't have any way of knowing whether it happened during the launch, in orbit, or during re-entry. If during launch, said experts Glanz consulted, it could be insulating foam; if during orbit it could have been hit by one of the 9,000 catalogued pieces of space debris the size of a softball already in orbit—or the countless numbers of smaller pieces, "which could be lethal to the Orbiter down to the size of a centimeter."

Excerpts from a news conference held on Friday by Mr. Dittemore, space shuttle program manager, revealed a discovery.

> We have recovered a partial wing leading edge RCC panel, that's reinforced carbon carbon. It's 26 to 27 inches long with 18 inches of wing structure still attached. We are still trying to determine whether or not it is the left or the right wing.

Reinforced carbon-carbon is a sheet of carbon which has been immersed in carbon resin. He was asked by reporters about the wiring bundle and replied they were working on that because there might have been one common event that caused the loss of sensors in many different locations. When asked about the foam, he said they were focusing on the foam loss and impact on the wing as a

> potential contributor. . . . It is our intent to test the foam and impact to a tile to prove that it either was a contributor or wasn't. . . . We're going to be proactive and prove that, by data and evidence, that it either was or was not a player—along with all the other fault tree branches and roots in other systems.

He added about the foam, "So it's still in our job jar." (*The empiricists, the testers, seem to have a chance to vindicate their position.*)

In our search for articles in the *New York Times* Archives, we turned up one essay in an area called the "Arts & Ideas/Cultural

Desk." Written by Edward Rothstein, the title is "CONNECTIONS; In Any Test of Human Limits, Death Is Among the Judges." The essayist tries to put the shuttle disaster in historical perspective. The essay opened with these three sentences: "Too much money, too much risk, too much death. In the week since the shuttle disaster, these have been some of the judgments offered as the space program has come under scrutiny. But when have such verdicts not been plausible for journeys of exploration?"

Rothstein recalled that in 1848 the British admiral John Franklin disappeared with two ships and 129 men while trying to find a passage through the Arctic. Thirty rescue missions failed with further loss of life; eleven years later the "wrecks of his ominously named ships, the *Erebus* and *Terror,* were finally found crushed by ice, the skeletons of the crew testifying to untold horrors and suffering."

This and other examples of death and horror Rothstein summarized were taken from a new book: *Dead Reckoning: Great Adventure Writing from the Golden Age of Exploration, 1800–1900,* edited by Helen Whybrow. These were accounts "of inhospitable worlds in which even the human is alien," not unlike the condition in space. These men were heroes, whose perilous travels were followed by the reading public in their first-person accounts of what they found.

Our age is much different from that Golden Age of Exploration. Rothstein thinks the space program created an almost democratic drama of space exploration. We stayed on the ground watching our chosen representatives performing in space, the International Space Station being a contemporaneous drama on a world scale. He credits Tom Wolfe and his book *The Right Stuff* for explaining this shift in explorer culture. In the days of *Sputnik* the American astronaut was a "cold warrior of the heavens," but to the test pilots they were not so much heroes as passive participants, "capsule-sitting observers." Risk and danger are to be avoided by NASA; society can tolerate it in safe dosages as packaged on television in the program *Survivor* and by athletes in armor participating in extreme sports.

Rothstein turns to the Lewis and Clark expeditions two centuries ago—he doesn't mention the irony that that they mapped the Columbia River—as treated in the book *Exploring Lewis and Clark,* by Thomas Slaughter. Exploration is not what we thought it was: "Exploring is a race with no second place." Rothstein summarizes the book in this way: The journals

from the Lewis and Clark expedition are full of distortions, claiming firsts when there were few, describing exaggerated dangers,
misunderstanding encounters with Indians, and ruthlessly con-
structing future reputations.

mythic heroes

Rothstein quotes Slaughter: "They already feared that they were
lesser men than the great explorers they emulated."
The author of the book being reviewed developed a counter
myth over and against the myth of the explorer as hero:

> . . . explorers almost universally resemble competitive imperial-
> ists, randomly killing animal life, blind to the visionary harmony
> of American Indian life, missionaries for a misguided culture. This
> is the familiar, postmodern vision of the Western explorer, a dis-
> torted inversion of its predecessor.

Some of this suspiciousness may have crept into the wariness about
space exploration, but what emerges again and again in the writings
of the nineteenth-century explorers is not the way in which they dis-
torted or destroyed or inflated but "the ways in which an explorer
comes to see." As a way of helping the reader decide between the
myth of the explorer and the countermyth, I reproduce Rothstein's
conclusion:

> The human is diminished in the face of immensity, vulnerable in
> the face of danger. The world, even in its horrific threats, can seem
> . . . "shining and white, possessed of all the purity we lack." This
> frailty, though, can also accompany an elevated vision of human
> possibility. Explorers survive to tell of their voyages. They test lim-
> its, at once submitting to them and defying them. It would be a
> mistake to retreat from such confrontations with limits. And as the
> disaster last week [*Columbia*] showed, it is impossible to avoid
> those limits completely. One of them has always been death.

As the chaotic information streamed in during the week after
the accident, it became clear that the mysteries were far from being
solved. NASA officials were puzzled by conflicting evidence. The
four different meanings of debris entered into the story—and they
had not yet been sorted out. The fault tree was still under construction. Chapter Four takes a break from the narration to provide a definition and explanation of organizational culture and other important communication concepts. It also takes a brief look at NASA's
history and its early culture. ✦

Chapter Three

Culture and Communication in NASA

The fact of the matter is that the "real world" is to a large extent unconsciously built up on the language habits of the group.

—*Edward Sapir*

"*Organizational culture is a communicatively structured, historically based system of assumptions, values, and interpretive frameworks that guide and constrain organizational members as they perform their organizational roles and confront the challenges of their environment*" (Modaff and DeWine 2002, 93, emphasis in original). The authors of this textbook definition go on to say that it is not as hot a topic as it was in the 1980s, when every manager seemed to have a book about organizational culture on his or her desk, but they take a normative stance, recommending cultural research by concluding that "a resurgence in organizational culture theory should occur" (103).

This chapter will attempt to contribute to the resurgence of the cultural analysis of organizations, or better, culture as communication, in this brief history and description of one organization. I could hardly avoid talking about culture in this case; the newspaper coverage and the very Board making the official investigation into the accident used the word cultural in a fairly consistent way.

Another textbook has a shorter definition of organizational culture: "*Culture is a system of meaning that guides the construction of reality in a social community*" (Cheney et al. 2004, 76). In addition to its brevity this one has the virtue of being a "constructive" definition; that is, the perception of reality is constructed by the culture. It also calls attention to a "system of meaning." It echoes to some extent the epigraph of this chapter, Sapir's idea that the world is unconsciously built up on the language habits of the group. If culture does work unconsciously on us to build up the world we perceive, then we should expect our cultural language habits to make certain parts of "reality" more or less visible.

Kenneth Burke once said that culture is a rough draft of human action that each of us writes, or enacts, in our individualized prose; I infer from the metaphor that *the stronger the culture, the closer our individual actions or drafts will be to the rough draft*. The differences between drafts can be attributed to *an individual's unique style of acting and talking and writing*. Language and meaning are thus the media of culture. Another word for language and meaning is *significance*. In his excellent book about interpreting culture, Clifford Geertz gave credit to Max Weber for the elegant metaphor he developed: that "man is an animal suspended in webs of significance he himself has spun" (Geertz 1973, 5); worthy of Spiderman, those webs of significance and meaning are the culture. These webs in which we are suspended are metaphors for the Whorf theory considered in Chapter Two. Webs of language—names—affect how we act in certain situations. I remind you of the unhappy example of the workers who were careless with matches around "empty" gasoline drums—drums that were full of dangerous vapors.

Redding's Ideal Managerial Climate

Closely related to the concept of culture is organizational climate. They are related concepts because managerial practices, or climate, have a lot to do with the total organizational culture. W. Charles Redding, regarded by most as the founder of the field of organizational communication, developed a model or ideal type by which to analyze actual communicative actions. He called it the "Ideal Managerial Climate" (1972). Redding didn't create the type out of his imagination, fertile though it was, but instead inferred it and the five factors from an encyclopedic knowledge of organizational research and theory that had been conducted up to the time of his synthesis. I summarize it briefly as follows:

1. *Supportiveness.* This word is used in its dictionary definition: furnishing support or aid. A supportive manager communicates with subordinates in a friendly, considerate, and helpful manner, recognizing in the deed of communicating the integrity of the other. Hierarchical differences are minimized by not "talking down" to the subordinate. The message received by the subordinate is that she or he has personal worth and importance.

2. *Participative Decision Making.* This concept denotes that people ought to be involved in the decisions that have importance for their work. Economist A. O. Hirschman's concept of "voice" in an organization is relevant (1970); a person with voice is one who is listened to, actively consulted, about important decisions. She or he is encouraged to express opinions, even *dissenting* ones. This implies that conflict is expected and desired. In the ultimate degree of participatory decision making, work groups or teams are given the authority to make collective decisions. Thus, organizational communication is far from being a one-way, downward-directed process.

3. *Trust, Confidence, and Credibility.* Redding treats these concepts as "close cousins" for good reason. Some models of communication consider credibility to be a dimension of interpersonal trust; others treat trust as a dimension of credibility. Experts on communication going back as far as Aristotle have said that source credibility is the most powerful factor in persuasion. I can't stress enough, however, that trust, confidence, and credibility are important in both sending and receiving messages. The receiver—whether a listener or reader—is more likely to be persuaded by a source perceived to be credible than one who is not. On the other hand, subordinates are more likely to talk about their problems with a boss they trust than with one they don't. Imagine how difficult it would be to tell your problems to a boss if you had heard he or she had used such knowledge against employees in the past. Underlying this factor is an *ethical*

base. People do not grant credibility and trust to superiors who lie; we do not trust those who have deceived us in the past. Effective communication is correlated, at least in the long run, with truth-telling.

4. *Openness and Candor.* Redding meant more than self-disclosure in selecting these terms. It is beneficial to communication when a source is open and candid. It can even increase her or his credibility, the third factor in the ideal managerial climate. It is equally important for receivers, particularly superiors, to be open to hearing dissenting opinions, even criticism. If not, the messengers bringing bad news may be "killed." When this kill-the-messenger attitude is present, external "whistle-blowing" is the only alternative for a member of an organization who observes something illicit, immoral, or illegal going on in it. While whistle-blowing is a courageous act, it almost always, even if undeservedly, brings retribution. We will look carefully at this, and the other four factors, in the case of the two shuttle accidents.

5. *Emphasis on High Performance Goals.* At first glance, this factor seems out of place, even irrelevant, to the communication process. Upon reflection, however, it does have a place in the ideal organizational culture or climate. By adding this factor, Redding avoided the trap of describing a "country-club" atmosphere, in which people chat with trust, credibility, candor, openness, equality—and nothing gets done. The competitive realities of the new global economy require hard work and high performance. Redding taught that the realization of the first four factors will make the fifth, or "bottom-line," criterion more attainable. Indeed, if an organization makes production or performance the only goal, and its managers lack supportiveness, credibility, and openness, then it will surely harm its employees and be riding for a fall.

These factors were drawn in part from a research area called superior-subordinate communication, one of the most heavily researched topics in organizational studies. Fredric Jablin's summaries of this research (1979, 1985) should be consulted by those more interested in the topic. How superiors communicate with subordinates determines the climate and culture of the organization. Supervisors spend between one-third and two-thirds of their time in communication with their subordinates. The attitudes they act out in this process have a major impact on people and organizations. Other topics listed by Jablin as important include openness, upward distortion of messages, upward influence (the ability of employees to persuade superiors about the need for change), semantic-information distance between superior and subordinate, feedback, and attitudes toward conflict.

NASA's Culture and Climate

The first point I must establish about NASA's culture was articulated in the standard book on the subject, Howard E. McCurdy's *Inside NASA: High Technology and Organizational Change in the U.S. Space Program* (1993). NASA never had one culture; rather, it has had from the beginning a confederation of cultures. This will become clear by explaining how NASA came into existence.

A convenient starting point in understanding the origins of NASA is October, 1957, when a small artificial sphere began to orbit the earth. The international situation at that time was, of course, the Cold War, a military standoff between the world's two superpowers, the United States and the Soviet Union. Each side tried to get ahead of the other in developing weapons of mass destruction, nuclear weapons and the vehicles to deliver them, including intercontinental ballistic missiles, submarines, and aircraft. Each side eventually built up enough weapons to destroy the other side, a stalemate called MAD, or Mutually Assured Destruction, to deter either side from actually using the weapons. In that context the Soviet Union suddenly achieved great international status—and caused acute embarrassment to the United States—by being the first power to launch an artificial satellite, in 1957, that orbited the earth, proudly beeping a signal for the world to hear. The Soviets called it *Sputnik 1.*

U.S. leaders quickly realized we were in a space race as well as an arms race with the Soviet Union. President Eisenhower and Congress created the National Aeronautics and Space Administration, NASA, the next year—1958—for the peaceful exploration of space and to catch up with the Soviets. Created is not quite the right word to describe the origin of NASA. Assembled is a better expression; a number of existing organizations were merged and some new ones were added to them (see Figure 3-1).

Each of the existing organizations had its own culture, some aspects of which were common to the other building blocks, some not. Let's look at some of the blocks that were put together. World War I had stimulated Congress in 1915 to establish a National Advisory Committee for Aeronautics, which, of course, became known by its acronym: NACA. In 1920 the Committee built a small research laboratory on an Army base near Hampton, Virginia: The Langley Memorial Aeronautical Laboratory. The purpose was to do research in flight engineering. The employees acted as, and thought of themselves as, researchers and engineers, not as government workers, and certainly not as bureaucrats.

The imminent outbreak of World War II (1939–1945) again caused Congress to act in this area by creating a new laboratory specializing in aerodynamics, particularly high-speed flight. The year was 1939. It was called the Ames Aeronautical Laboratory at Moffett Field in California. It soon created a strong engineering culture.

In 1940 Congress created another laboratory near the municipal airport in Cleveland, Ohio: The Lewis Flight Propulsion Laboratory. It did basic research on aircraft engines, and the engineers were additionally motivated by the knowledge gained from secret reports that our eventual enemy, Nazi Germany, was developing some revolutionary aircraft engines.

NACA now had three labs and a small administrative center in Washington, D.C., staffed by fewer than 200 people. But the three centers had a proud independence necessary for the culture of research laboratories. They were the premier institutions in the United States for aeronautical and engine research to hasten the development and production of military airplanes. Another dimension of the developing culture emerged when the Langley engineers set up a field station called the High Speed Flight Research Station in the Mojave Desert of Southern California. They later developed the Bell XS-1 airplane, and several test pilots were killed while trying to fly prototypes. An Air Force test pilot named Charles E. "Chuck" Yeager would eventually break through the sound barrier in the experimental plane. He and the other daring, fearless test pilots would be described by writer Tom Wolfe in his book *The Right*

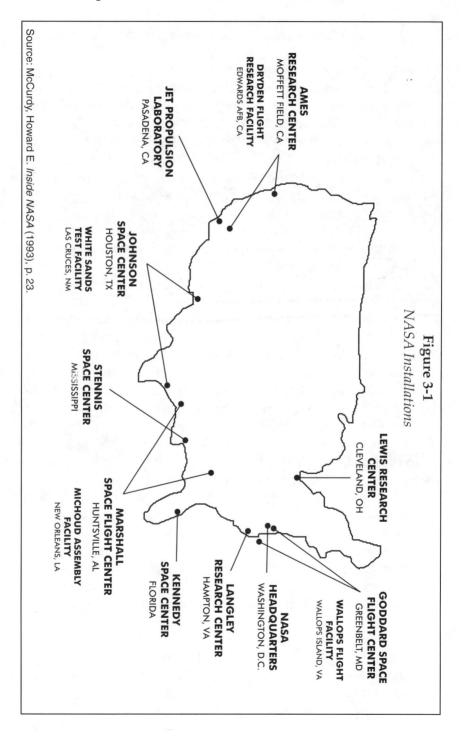

Figure 3-1
NASA Installations

Source: McCurdy, Howard E. *Inside NASA* (1993), p. 23.

AMES RESEARCH CENTER
MOFFETT FIELD, CA

DRYDEN FLIGHT RESEARCH FACILITY
EDWARDS AFB, CA

JET PROPULSION LABORATORY
PASADENA, CA

JOHNSON SPACE CENTER
HOUSTON, TX

WHITE SANDS TEST FACILITY
LAS CRUCES, NM

STENNIS SPACE CENTER
MISSISSIPPI

LEWIS RESEARCH CENTER
CLEVELAND, OH

MARSHALL SPACE FLIGHT CENTER
HUNTSVILLE, AL

MICHOUD ASSEMBLY FACILITY
NEW ORLEANS, LA

KENNEDY SPACE CENTER
FLORIDA

LANGLEY RESEARCH CENTER
HAMPTON, VA

NASA HEADQUARTERS
WASHINGTON, D.C.

WALLOPS FLIGHT FACILITY
WALLOPS ISLAND, VA

GODDARD SPACE FLIGHT CENTER
GREENBELT, MD

Stuff; a motion picture about these brave men appeared with the same title.

NACA had not developed any kind of craft to explore space. Rockets and missiles were regarded as weapons, and as such their development was logically assigned to the three branches of military service, the Air Force, Army, and Navy. The Army's rockets were developed by the ABMA, the Army Ballistic Missile Agency, at the Redstone Arsenal in Huntsville, Alabama. Their approach was called the "arsenal concept" because they believed in having an in-house, in-depth technical capability of engineers to do the research and development on rockets and missiles. They relied on aerospace contractors in the private sphere to a limited extent, choosing to fabricate—that is, construct—and test the rockets and their components; contractors would then produce them in large numbers under the close supervision of Redstone Arsenal engineers.

The Air Force was late getting into the rocket business, choosing instead to develop long-range bombers capable of flying over the territory of the Soviet Union and their allies to drop bombs. The Air Force also rejected the arsenal concept of doing research and development in-house, electing to rely on aerospace contractors to do the research and development (R&D), as well as the fabrication, of the craft. They even used contractors to evaluate the bids and proposals of other contractors. For these and other reasons the Air Force was not in the race to be the first branch of the service to get into space. The race was between the Army and Navy.

Two years before *Sputnik,* in 1955, the Defense Department tapped the Navy to launch the first American satellite. The project, called Vanguard, was carried out with the Navy's version of the arsenal concept at an in-house facility, and was supposed to put the satellite in orbit during the International Geophysical Year of 1957–1958. On December 6, 1957, the Navy attempted its first launch of its Vanguard Program. McCurdy said the launch would have been "comical" if it had not had such serious implications for the Cold War. The rocket flew about three feet in the air, paused, and fell apart, the "three-pound satellite rolled away, beeping pointlessly" (16). The Soviet Union had launched *Sputnik 2* a month after *Sputnik 1,* a frightening warning that they had a clear technological lead in space, accompanied by the possibility that this could be translated into a military advantage over the United States and its allies, including spy satellites and weapons launched from space.

President Eisenhower turned to the Army. Within three months the Army successfully orbited *Explorer 1*, a small scientific satellite. The rocket that put it in space was called Juno, a rocket developed by the ABMA at the Redstone Arsenal in Huntsville, Alabama. The leadership of this group was unique. The Director was Dr. Wernher von Braun, the technical wizard who had developed the V-2 Rocket in his native Germany during World War II. He and 120 other Germans had surrendered to the U.S. forces at the end of the war. They were brought to the United States in what was called Project Paperclip and went to work for the U.S. Army to develop much of the American Cold War arsenal of guided missiles.

On October 1 of 1958 the three NACA labs became part of NASA. Other installations, other building blocks, were needed, however, in order to assemble the new agency and enter the space race. The rocket engineers at the Redstone Arsenal were transferred from the Army to NASA in 1960. They wouldn't move far in a geographic sense because NASA would build the George C. Marshall Space Flight Center for them outside the city limits of Huntsville, Alabama. About 4,500 employees would move, under the leadership of von Braun and the other Germans, to the new space center. They would retain the arsenal concept they practiced in the Army even though they were now members of a civilian agency.

There was another missing block—a space center to do research and development into the human factor, or human space flight. In the beginning this was handled by the Space Task Group at the Langley Research Center; the NACA lab' shifted from NACA to NASA with the assumption that its activities would be moved to a new NASA field center. In 1962 the Space Task Group was moved to a new facility south of Houston, Texas, with what would over time become a somewhat inaccurate if not sexist name: the Manned Space Center; it would be renamed the Lyndon Baines Johnson Space Center in 1973. The Johnson Space Center, or JSC, would train the astronauts for space flight, design space capsules and spaceships, and direct operations on all flights with human beings from its Mission Control Center near Houston.

JSC would develop a culture that produced tension and competition between it and the Marshall Space Flight Center, or MSFC, in Alabama. JSC, for example, didn't completely accept the arsenal concept, having fewer engineers and relying more heavily on private contractors. They also wanted to have control over all spacecraft in which astronauts would fly, an ambition shared by the Germans in Alabama. Von Braun had sketched an early version of the moon rocket in the early 1940s and had in the early 1950s proposed to build a space station with a winged space plane that could service it. The German group didn't want to be limited to building rocket engines and boosters.

There was one more missing block. If you had building blocks where you could design, build, and test rocket engines (MSFC in Huntsville) and a place where you could train the astronauts, design their spacecrafts, and control the missions from earth (JSC in Houston), where would you launch them? The Germans had from the beginning launched their rockets in Germany and once in the United States, at White Sands, New Mexico; they had the earliest and most extensive experience launching missiles in the U.S. complex. The Air Force had a launch facility in Florida, but NASA didn't want to be under their control, so they pushed for an all-NASA launch complex. NASA thus came to build the John F. Kennedy Space Center at Cape Canaveral, Florida. A member of the German team, Kurt Debus, a technical assistant to von Braun, was sent to be the first Director of the new launch facility. That would guarantee a close cultural connection and similarity between the Kennedy Space Center, or KSC, and the Marshall Center.

NASA also established other centers around the country devoted to specific purposes, some of them in support of the other centers, but the main building blocks have been identified. The assembly of existing organizations produced a confederation of cultures. And although I will attempt to speak of a NASA culture, the reader should keep in mind the differing cultures of the different NASA centers. McCurdy supplies the analogy of space centers and universities: "In many ways, the separate centers behave like rival universities, each with its own traditions and interests" (23). Well, they don't have volleyball teams and debate teams, football teams and cheerleaders, but there is something to the analogy. And yet rival universities have more in common with each other than they have with private businesses or government bureaucracies. They are all, for example, committed to the truth, to research and education, even if they compete with each other for football victories and financial resources. So it is with NASA.

The next step is to describe the NASA culture, a difficult job with any organization, but made easier for us by the work of Howard E. McCurdy. McCurdy approached his task in several ways, by searching archives, by interviewing key NASA employees, and by a questionnaire he sent to a sample of employees in 1988. Following are the main points of his description of the NASA culture, with commentary based on my experience of the Original Technical Culture in the late sixties.

Cultural analysis

Research and Testing

The original NACA culture placed weight on research into aeronautics and flight. Instead of centers, their installations were called laboratories. The NACA employees turned out more than 16,000 research papers, just like faculty members at research universities. They were on the leading edge of aeronautical studies. McCurdy stresses that NACA developed a technical culture around the values of research that was unlike the bureaucratic culture of other government agencies. NACA employees had a passion for truth, rather than politics. In fact, we might label this first element of culture "seeking the truth through research and testing." Testing and verification is the road to truth in any kind of research institution; at least, that was the original assumption.

The new space centers of NASA were just as committed to research and testing as were the three original laboratories of NACA. Components to engines and space ships were tested and retested; then the components were put in the engines and space ships where both entities were tested and retested. The engineers learned from testing, learned what worked and what didn't. What didn't work 10 out of 10 times was redesigned, tested, and retested. The Germans under von Braun were known as "incrementalists" because they would start with the smallest details, perhaps a relay, put the parts in a circle and test them until they knew they were reliable. Then they would put the relay in the larger unit to see how it would behave under tests. The rockets themselves were first launched, not to prove the engineers were smart, but to see how they would perform, to find what changes would have to be made before putting astronauts on top of them.

The Marshall Center had its own test facilities, giant shake stands that could torture a rocket in a way duplicating the stresses of actual space flight. Giant test stands gripped rocket engines so that they could be fired without producing an unscheduled launch from Alabama. When the contractor responsible for building and testing them would deliver them to the Marshall Center, the engineers would put them through their own tests. This usually irritated the contractors, making them feel as though they weren't trusted. Well, they weren't.

One difference between the original NACA labs and the NASA space centers, such as MSFC, was the attitude toward producing scientific papers and reports. As stated earlier, NACA personnel believed in the practice of writing papers, just like the culture at research universities in which the professors must "publish or perish." Dr. von Braun, by contrast, said he began to worry if he saw that his engineers were turning out too many papers and reports. The culture at the Marshall Center demanded that the knowledge learned from research and testing had to be applied to the hardware, the rocket engine or the space capsule.

move from papers → applied research

In-House Technical Capability

When NASA was assembled and created in the late 1950s and early 1960s, it was assumed that the agency would need to have well-educated technical people to do the job. Although McCurdy does not emphasize the relationships among cultural elements in his analysis of the early NASA culture, the second element of the culture goes with the first, research and testing. How could the agency go about the search for truth by research, testing, and verification without well-trained engineers?

In-house technical capability is a basic element of the arsenal concept the Marshall Center brought with them from their experience in the Army Ballistic Missile Agency (ABMA) at the Redstone Arsenal. As an example, the Redstone missile was an important part of America's defense program; it was designed, tested, manufactured, and retested at the arsenal. The first 17 missiles were fabricated at the arsenal. Only after the arsenal had accumulated that much experience and knowledge was the contractor, Chrysler, allowed to manufacture them in large numbers. Even then the ABMA engineers monitored Chrysler closely.

An in-house capability also promoted flexibility in the centers. If something went wrong and the facility didn't have technical muscles and tools, it couldn't be fixed. Testing and verification go with—indeed facilitate—the notion of fixing things that go wrong. If you had to rely on a contractor, you would have to make a drawing for them to understand the problem and then wait for the solution—which might or might not work. With the in-house depth of technical talent, scientists and engineers were flexible, able to make a correction within minutes, test the correction, and move on. It's a bit like having an expert on computers or electricity or plumbing in the family, in-house so to speak.

Von Braun had a rule during the Apollo Program: at least 10 percent of the money spent on rockets must be spent in-house. On the basis of my experience working for him, I can verify that he and his associates regretted the 90 percent that got away. They wanted the money to pay the people to do the design and testing, and to recruit new and younger talent. There was resentment when money went to contractor personnel who displaced able NASA people. They felt the system was strengthened when MSFC had technical expertise *in depth,* the better to monitor the contractors who, after all, had to make a profit.

Hands-On Experience

NASA found that the opportunity for hands-on experience was an effective recruiting tool; it was a powerful incentive to offer young recruits. Young people trained at the top engineering schools in the country wanted to put their hands on the important tools and problems, thereby improving their engineering skills and increasing their knowledge experientially. I suspect that most if not all college students share the desire to try to apply what they've learned in classes. Even the Johnson Space Center in Houston, reliant as it was on private contractors, used NASA employees at the consoles in Mission Control. Astronauts were NASA employees. Engineers didn't want to spend all their time reading contracts and monitoring the private contractors; they wanted to be designing the hardware, fabricating rockets, and testing their designs in rigorous ways.

Most of the engineers didn't want to become managers. I discovered a management problem at the Marshall Center during my research there in 1967 and 1968 as a Summer Faculty Consultant in organizational communication. The MSFC culture hammered people every day with the idea that an engineer has to keep his hands dirty, another way of saying keeping their hands on the hardware, not doing the paper shuffling. This value was so deeply inculcated in the employees in the laboratories that they didn't want to become supervisors or managers. I called this (Tompkins 1977; 1978; 1993) the problem of the "Reluctant Supervisor." Supervisors had to shuffle papers and supervise people who were getting their hands dirty. The only incentive to become a supervisor was an increase in pay. So the engineers accepted a promotion, and an increase in salary, to the level of supervisor with reluctance. This hampered their ability to do the job effectively as a supervisor, longing as they did to get back in the lab and put their hands on the technology.

I found that this value created a division or conflict within the Marshall Center. The Apollo Program was so immense—25 billion in 1960s dollars—that contractors had to do 90 percent of the work, or at least they got 90 percent of the money. But NASA and MSFC didn't want to give up the direction or control of the programs. The ten or twelve laboratories remained similar to the ones established in Germany to do the research and development of the V-2 Rocket. But the Saturn V, the moon rocket, was so large and complicated that MSFC created a new branch of the organization called Industrial Operations, or IO. The laboratories got a new title, Research and Development Operations, or RDO.

Managers and staff members in IO tried to coordinate the work between RDO and the private contractors by means of a new form of organization, the matrix or matrix overlay, in which horizontal and diagonal channels of communication were nearly as important as vertical ones. Where would you find the managers to do such a job of coordinating the work among people over whom they didn't have organizational authority? They came out of RDO, in some cases "recommended away" by lab directors who wanted to replace them. The engineers in the RDO labs were still getting their hands dirty, while the new managers, called project and program managers, shuffled papers, talked to contractors, and tried to get their former colleagues in the labs to help them. The engineers resented this new interference and soon there developed a division between RDO and IO, what I called a "lack of lateral openness" (1993, 88–90). This was a serious communication problem: a cross feed of communication was required between IO and RDO, a problem that von Braun gave considerable attention to—a problem created at least in part by the belief that NASA engineers had to keep their hands dirty.

Exceptional People

By definition, most people working in U.S. organizations and institutions are average. There are exceptional people, outliers, above and below the average in intelligence; that is the nature of the bell-shaped curve. Indeed, in a later chapter we shall see, ironically, that there is a controversy among students of organizations as to whether smart people are always good for organizations. Setting that question aside for the moment, it is obvious that the original technical culture (OTC) of NASA assumed the need to recruit and keep the best and brightest from U.S. colleges and universities.

Folklore since the Apollo Project reflects the belief that NASA did recruit exceptional people: "You don't have to be a rocket scientist to understand . . ."; and, "If we can go to the moon why can't we solve the problems of. . . ." Personal experience confirmed this element for me; the NASA employees of the original technical culture were indeed exceptional, starting with von Braun himself, a Ph.D. in physics by the age of 22, pilot, polyglot, mountain climber, musician, and composer. McCurdy not only makes his point about how gifted and well educated NASA workers were in the OTC, he asks how the organization could have attracted them. Attract them NASA did, inducing top college graduates in all the engineering disciplines, as well as experienced workers from other government agencies, universities, and private corporations to join the space race. *(The German Team, of course, didn't have to be attracted.)*

How they did it is a good question, considering that NASA salaries were well below those offered by the aerospace corporations for the same or similar work. McCurdy's answer is this: "What motivated people to join NASA was the challenge of the work" (1993, 53). That statement is true but there is more that can be said of the attraction. McCurdy doesn't seem to be aware of scholarly work in the area of organizational identification. Its importance was recently expressed by Daniel P. Modaff and Sue DeWine (2002) in their book, *Organizational Communication: Foundations, Challenges, and Misunderstandings*. In the introduction to their book they identify three important constructs, the first of which is organizational identification. (The second and third are job satisfaction and communication satisfaction.)

Coincidentally, the impetus for my theorizing about and studying organizational identification came from my experience at the Marshall Space Flight Center. Reflecting on my feelings and experiences as I began my long drive home from Huntsville to Detroit, I realized that I had entered a state of *identification* with the organization. I felt a part of it, a sense of belonging, and had persuaded myself that we had common interests: What was good for the space program was good for me. There was an *inherent emotional component*, positive in nature, a symbolic satisfaction in my relationship with the organization (similar to what I had observed about NASA personnel). I also experienced a change of identity as a result. The organization became part of me; I mentioned my involvement in the space program in conversations with friends and acquaintances and in my lectures to students. Nonetheless, it wasn't a blind loyalty. I felt an automatic responsibility to identify organizational problems at NASA and try to solve them. That is what kindled my theoretical and empirical interest in the concept of organizational identification.

A description of the subsequent development of theory and the research program in identification and its unobtrusive control over people is taken up in a later chapter, but notice for now the parenthetic expression in the paragraph above in which I said my experience was similar to what I had observed about NASA personnel. They worked for much less money than they could have made elsewhere not only for the challenge of the work but also because of their identification with the organization and its people, aims, and projects. Identification helps explain organizational motivation. We can now add, with some confidence, organizational identification to McCurdy's list of OTC factors.

Risk and Failure

As exceptional as the original NASA members were, they knew they were bound to fail part of the time. That is the nature of space flight. Astronauts and earth-bound engineers and technicians, secretaries and staff members, managers and faculty consultants, all of them understood the risks inherent to space travel, particularly with humans aboard the ships. Launches that worked were occa-

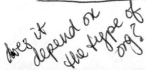
random successes ↗ & morale?

sionally referred to as "random successes." It is much safer to travel on a commercial airliner, of course, than a Saturn V, the moon rocket, or a space shuttle.

Part of the legend was due to the belief, says McCurdy, that one of four test pilots was killed in experimental aircraft trying to break the sound barrier. The Mercury astronauts flew aboard an Atlas intercontinental ballistic missile; the Air Force had them built with less reliability than vehicles to take people into space because it was cheaper to order extra missiles and assume some would fail. The first time it was tested as a space vehicle it failed; fortunately, no astronauts were aboard at the time.

Under this heading of risk and failure McCurdy presents a remarkable sentence: "The normalization of risk, the acceptance of failure, and the anticipation of trouble led to an atmosphere in which these things could be discussed openly" (1993, 65). The sentence is remarkable for its layers and segments of meaning: (1) the phrase "normalization of risk" is similar to a phrase that would grow out of a later analysis of the *Challenger* disaster, "the normalization of deviance," setting up a comparison and contrast for later chapters of this book; (2) the phrase "acceptance of failure" can be read as a cynical as well as a realistic interpretation of the culture; (3) the phrase "the anticipation of trouble" evokes the "earthquake prediction" metaphor that von Braun expressed to me, which will be discussed in Chapter Four; and (4) the phrase "an atmosphere in which these things could be addressed openly" introduces a brief discussion of the importance of organizational communication in the original technical culture of NASA.

It is imperative at this point to unpack the final phrase, the notion that risk and failure created an atmosphere in which these things could be discussed openly. McCurdy presents several pages in which interviewees are quoted about how things got talked about and solved, a lack of tolerance for the "yes man," and a "great deal of democracy" in the early years of NASA, and then this generalization: "Open communication was very much a part of the early NASA culture" (1993, 65). And yet it is discussed under the heading of Risk and Failure. McCurdy's own justification of open communication as an important part of the culture makes it more than a subhead; with considerable justification open communication can be added as a dimension of organizational culture. One can also ques-

does it depend on the type of the org?

tion whether open communication is "part" of the culture, or constitutive of it.

Frontier Mentality

The metaphor of space as the last frontier for Americans still has
some power, even though it was more applicable to the first decade
of NASA's life than it is now, with the repetitious, yet dangerous,
shuttle flights. McCurdy reminds us that in the early days, the
agency moved from challenge to challenge. After the Mercury Program was completed they moved on to Gemini; next came the
Apollo Program, the flights to the moon, and then the Apollo Applications Program, including Skylab.

> After three Mariner spacecraft flew by the planet Mars, *Mariner 9*
> went into a Martian orbit. After *Mariner 9* mapped the whole
> planet, NASA officials sent *Viking 1* and *Viking 2* to the surface of
> Mars to look for life. It was remembered as a golden era by those
> who worked for NASA. (McCurdy 1993, 72)

In summary, let's enumerate the elements or parts of the early
NASA culture, adding two new elements to McCurdy's headings:

1. Research and testing

2. In-house technical capability

3. Hands-on experience

4. Exceptional people

5. Risk and failure

6. Frontier mentality

7. Open communication

8. Organizational identification

We should now return to the definition of organizational culture
with which we began the chapter. Do these eight elements fit the
definition? They fit the defining characteristics of assumptions and
values and interpretive frameworks. The one element out of place is
number seven, "open communication," because the definition
makes communication a "constitutive" factor of the other elements
of culture; that is, communication is necessary to persuade people
of the need for research and testing, to having in-house technical ca-

pacity, to hands-on experience, to recruit exceptional people, to bring people to the acceptance of risk and failure, and for open communication, a qualified kind of communication. Communication in general is, thus, an overarching dimension of culture, and yet we can say communication is constitutive of the sub-genre, communication qualified as open communication. We can, therefore, logically retain number seven in the list and make all eight elements subheadings under the general heading "Communication and Culture at NASA," which were virtually synonymous during the old technical culture. That will be the argument of Chapter Four.

Another qualification is that McCurdy left out one factor that most anthropologists would consider essential to any understanding of culture—language and meaning. We saw in the previous chapter that the Sapir-Whorf Hypothesis was developed by an engineer looking for the causes of catastrophic fires. He found some of them to have their origins in the language of the group, the work group. Engineers have a language all their own. Dr. von Braun, while sending me out to look for communication problems in the organization, warned me to attend to what he called the "alphabet soup" of acronyms the engineers used, utterances foreign to his native nomenclature of physics. We shall try to be attentive to possible problems of language and jargon as affecting meaning and the behavior of members of the culture. We shall consider it under the heading of communication.

Before moving on, however, we must address an important question posed by Eric M. Eisenberg and Patricia Riley in their learned work about organizational culture. Summarizing the work of others, they wrote that "so-called strong cultures do not necessarily result in excellent performance and can even be destructive unless included among their norms and values is a focus on adaptation to a changing environment" (Eisenberg and Riley 2001, 310). I shall provide a tentative answer to the question implicit in their statement: What communicative practices could provide a strong culture with the ability to adapt to a changing environment? The early NASA culture we have described is surely a "strong" culture. Complicating the question is the fact that the strength of the culture did not remain a constant. We can, therefore, describe what happens when a strong culture changes.

In Chapter Four I shall move more deeply into the organization called NASA during the golden era, examining the Marshall Space

Flight Center in Huntsville, Alabama. It was the largest and perhaps most important NASA center at the time I entered it as a Summer Faculty Consultant in 1967. We shall examine communication-as-culture by moving down the hierarchy from the Director's office to the floor of the engineering laboratories. ✦

Chapter Four

Communication and Culture in the Marshall Space Flight Center

In Chapter Three I presented, and amended, the aspects of the original technical culture (OTC) of NASA as articulated by McCurdy (1993). While McCurdy was writing his book, I was writing one about the Marshall Space Flight Center (Tompkins, 1993) and we were unaware of each other's work. This chapter will present first-hand observations and data gathered by me in 1967 and 1968 as a Summer Faculty Consultant in Organizational Communication at the Marshall Center. There are two purposes in this chapter. The first is to make a deeper description of culture and communication than in the previous chapter about NASA as a whole. The second is to provide additional evidence in support of McCurdy's claims and support for the amendments suggested in Chapter Three. That will make it possible to look for changes in the culture of NASA in the report to be issued by the Columbia Accident Investigation Board (CAIB).

A Research Opportunity at MSFC

As I wrote in the Preface, I was recruited to work for NASA by Walter Wiesman, then coordinator of internal communication at NASA's Marshall Space Flight Center in Huntsville, Alabama. He had obtained my name from my Ph.D. adviser and former colleague at Purdue University, W. Charles Redding, and asked whether I'd be interested in spending the summer as a Summer Faculty Consultant in Organizational Communication at the Marshall Center.

I decided to accept the offer even though it took five months to fill out all the papers and get a security clearance from the FBI. As I drove from Detroit to Huntsville, I was alternately depressed by radio news about the war in Vietnam and elated at the prospect of joining the Apollo Project. President John F. Kennedy had stirred me, as well as most Americans, with his special message to Congress on May 25, 1961: "I believe that this nation should commit itself to achieving the goal, before this decade is out, of landing a man on the moon and returning him safely to Earth." Heading south, I realized we had only two and a half years left in which to fulfill Kennedy's commitment.

I had little idea of what I was getting into at the Marshall Space Flight Center (also known as MSFC). I did know that the director of MSFC was the legendary Wernher von Braun and that the mission of the Center was rockets and propulsion. In 1967 its main concern was the Saturn V, the mighty moon rocket that, if successful, would enable the United States to overtake the Soviet Union in the space race. Ever since the Soviet Union had demonstrated its apparent superiority in space by launching *Sputnik 1*, 10 years earlier in 1957, the United States had suffered from a technological inferiority complex, intensified by the fervent anti-communist sentiments of the period.

June 26, 1967, was my first day at the Space Center. I was part of what was then called the Manpower Office, working under Wiesman. He was the youngest—"the kid"—of the German rocket team, the 120 experts who von Braun had brought with him to Texas during "Project Paperclip" (the code name for the U.S. Army's interrogation and recruitment of the team after World War II). Wiesman was one of the few nontechnical members of the team. From his reading and observations, he had concluded, like von Braun, that communication was the essence of organization and management. He joined several professional and academic societies, met W. Charles Redding, who had directed my doctoral dissertation five years earlier and who in turn referred me to Wiesman.

On my second day of work, I attended a meeting of the Advanced Systems Operation, an internal think tank devoted to planning space flight as far into the future as 1992. (Some members of the group, I was told, served as consultants to Stanley Kubrick during the filming of *2001: A Space Odyssey*.) Wiesman spoke about the importance of communication. They had several items about communication on their agenda. This was part of my orientation to Wiesman's program and the Center, where 7,200 people were working on the Saturn V and other projects.

As part of my orientation, Walt Wiesman arranged several tours of the huge campuslike Space Center for me. I visited the design and manufacturing labs and "clean" rooms, and watched astronauts simulate neutral gravity with scuba gear and weights in a tank of water. Rockets and boosters were tested in ingenious ways to imitate the tortures of space flight. Giant shake tables, like huge backyard swings with locomotive drive rods, mimicked the stresses of a launch. Most exciting was the test of the mighty F-1 engine (five of which were clustered at the base of the Saturn V moon rocket). I was shown into a concrete bunker where we watched the test through a slit. Outside, on a patch of red Alabama dirt, sat the powerful engine, held in place by a towering test stand. The rocket blast would roll off a special metal deflector at a 35-degree angle. Cold water began to run down the shields and deflector during the ignition countdown. My guide, an engineer, explained that this would prevent the test stand from melting. "Think of a rocket engine as a kind of controlled and continuous explosion," he told me.

The explosion was the loudest sound I ever hope to hear: 1,500,000 pounds of thrust from one rocket engine. As the flames hit the water, a vast cloud of steam engulfed the test stand; enormous clouds of smoke roiled skyward. Trees several hundred yards away bent over backwards from the shock waves. Then the roar abruptly ceased.

"Why didn't the engine launch the test stand?" I asked in the uncanny silence.

My guide answered, "It's the difference between starting to tow a car with a chain that's either slack or taut. We don't allow the engines to lunge at the test stand."

The test stand and shake table and other equipment were used in the research stage to find problems and fix them. They were also used as a kind of verification after rockets and engines were manufactured, whether by MSFC or the contractors. McCurdy's observation about research and testing as a part of the early NASA culture is correct as it applied to the Marshall Center, the biggest and most important of NASA's field centers at the time.

As part of my orientation to MSFC, I was sent to New Orleans, Louisiana, to visit the Michoud Assembly Facility; in those days it housed the Boeing employees who built the Saturn V, the moon rocket, under the supervision of Marshall Center managers, engineers, and inspectors. (Today the facility is used to make the 15-story external fuel tank, complete with the insulating foam, for the space shuttle.) While there I took a NASA bus to the Mississippi Test Facility (MTF), an extension of the Marshall Center. (Today it is a separate center, the John C. Stennis Space Center, named after a Mississippi Senator who was supportive of the space program.) The Marshall Center did so much testing of its rocket that it needed more room. MTF was in Hancock County, Mississippi, the location being selected because of the virtual absence of human beings. When they cleared the trees and vegetation, however, they found a vast complex of illegal stills that had been used to make, ironically, *moonshine.* The huge area was used for testing, testing, and retesting, additional confirmation that research and testing were major aspects of the MSFC culture as well as of the NASA culture.

The Arsenal Concept

As I developed a better grasp of the organization, I learned about the history and culture of NASA and how it dealt with contractors. Early on, NASA had to decide whether to use the Army approach or the Air Force approach, two radically different modes of operation. Under its arsenal concept of developing weapons, the Army had maintained an in-house technical capability that permitted it to conduct its own research and development (R&D), including the capability of manufacturing weapon prototypes. This experience gave the Army a yardstick, a metric, so to speak, by which to measure the quality and cost of the items contractors delivered to them.

The Air Force, by contrast, maintained little technical expertise, relying instead on private contractors to propose new weapons, conduct the R&D, and manufacture the planes and missiles. As a consequence, the Air Force lacked the means to assure quality control of what it was buying and had limited ability to monitor costs. This approach, however, was politically effective with Congress for the reason that it had a corps of contractors around the country and the more of them there were, the larger the lobby for their projects. According to what I was learning by listening and from a study by Nieburg (1966), such a system generated a large, geographically diffuse team of senators and representatives to vote for higher military appropriations, which, in turn, were translated by contractors and payrolls into voters in their states and legislative districts.

When James E. Webb became Administrator of NASA in 1961, he wanted to eschew the arsenal concept in favor of the Air Force's contract system. According to Nieburg (1966), NASA sought to dismantle the Marshall Center's in-house capability. MSFC, fearing the consequences of breaking its yardstick, resisted this attempt mightily, and the ensuing struggle produced a long-term tension between MSFC and NASA headquarters. Von Braun and his colleagues wanted not only to do the R&D for the Saturn V, they would have preferred to *build* the moon rocket at MSFC. NASA overruled them, beginning a trend that would continue for the next several decades in which the Marshall Center shrank and the contractor corps grew.

Organizational Structure at MSFC: Mitosis

In Germany, in the U.S. Army, and initially at NASA, von Braun had acted as chief engineer, supervising the efforts of ten or more laboratories during the R&D and initial manufacturing stages. But as the tasks multiplied with the creation of the Apollo Program and the goal of sending men to the moon, the team had to change its organizational structure. The Mercury and Gemini programs had used upgraded Redstone missiles, but Apollo required the creation of new and bigger rocket technologies. This led to the development of the Saturn series, beginning with Saturn I and IB and culminating in Saturn V. The organization had to grow, and it did so by mitosis (a form of metamorphosis by which a living cell divides into two separate entities). This growth period for MSFC followed the pattern described by Barnard (1938) of all complex organizations:

> [All] organizations of complex character grow out of small, simple organizations. It is impossible for formal organizations to grow except by the process of combining unit organizations already existing, or the creation of new units of organization. (104)

The muscle of the MSFC organization was supplied by its laboratories. They were retained in much the same form they had taken at Peenemünde, Germany, during World War II, under the organizational title of Research and Development Operations (RDO). RDO was organized along the lines of various science and engineering disciplines. There were 12 laboratories in 1967: the Advanced Systems Office (futurists); Systems Engineering (integrators); the Aero-Astrodynamics Lab (trajectories); the Astrionics Lab (electrical engineering); the Computational Lab (computer science and service); the Manufacturing Engineering Lab (prototype builders); Propulsion and Vehicle Engineering (rocket motors); the Test Lab; the Space Sciences Lab (astrophysics); the Quality and Reliability Assurance Lab (inspectors for quality control); and two small units, Experiments and Operations Management. Notice the cross checks built into RDO itself, what with people to build rockets, people to inspect them, people to test them, and of course the specialized disciplines. Personnel in RDO continued to "keep their hands dirty," a high value in the culture, but now they were working alongside thousands of people employed by private contractors.

In the mitosis of MSFC, the new division of the organization created for the Apollo Program and to accommodate NASA headquarters, was called Industrial Operations (IO). The chief responsibility of IO was to direct and monitor the efforts of prime contractors. There were over 100,000 people in the public and private sectors involved in the Apollo Project, the largest engineering project in history. IO was organized into offices relating to the various programs and projects for which the Center was responsible. The Saturn V Program Office, for example, monitored R&D and manufacturing for the huge rocket. It was further divided into project offices, each specializing in the various stages and engines of the Saturn V, thereby producing a highly matrixed organization.

Project Management

The theory of project management, new at the time, was that managers and other personnel could shift from office to office as projects and programs came and went. Each rocket was managed by a matrix of offices, one concentrating on the whole, others on individual rocket stages and engines. This created a part-whole relationship and a certain degree of redundancy. Project managers in IO could ask for technical help from the labs in RDO as well as from contractors they were monitoring. Von Braun described the changes

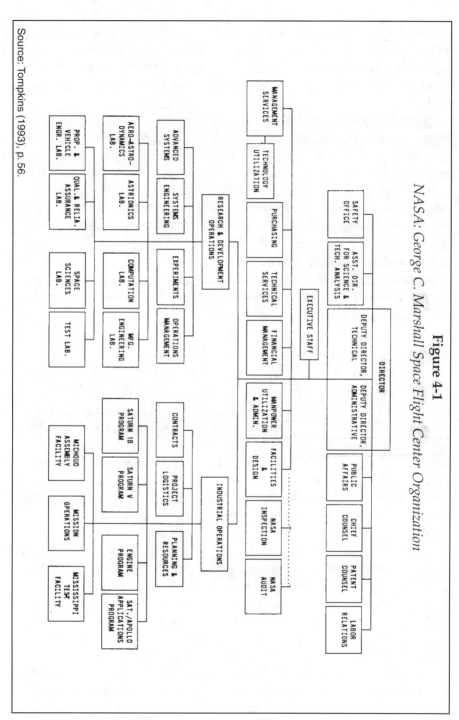

Source: Tompkins (1993), p. 56.

Figure 4-1

NASA: George C. Marshall Space Flight Center Organization

during my first interview with him: "We went through a metamorphosis into a management organization—monitoring and detecting. We went from a do-it organization to see-that-it-is-done job. We had to hire many new people and new occupations, such as accountants."

Both RDO and IO reported to their respective directors, who, in turn, reported to the director of the Center, von Braun. For the first time in his experience, there was a level or layer of management between von Braun and the laboratory directors. He was assisted by two deputy directors, one for technical matters, and the other for administrative affairs. Various staff offices were also connected to the director's office. After mastering the organization chart and visiting most of the offices and labs, I had a good grasp of the structure of the Center. I would soon come to understand the dynamics of the organization, hidden under the organization chart, by sitting in on technical briefings for von Braun and from formal interviews with the top 50 managers and lab directors.

Formal Versus Informal Organization

The preceding description is of the formal organization of MSFC. The formal organization is rational; lines of authority are consciously drawn. The lines on the organization chart are formal channels of communication; they describe who can give orders to whom and who reports to whom. It is largely impersonal. The informal organization is quite different. It is based on personal, as opposed to organizational, interactions. People working in an organization develop personal relationships. These relationships might develop in a company bowling or softball team, in meetings at a cafeteria, in car pools, or in other chance meetings.

Scholars of communication assumed that the formal organization gave rise to informal communication and informal organization. Barnard (1938) first discovered that it also works the other way: Informal communication and organization give rise to the formal organization. Suppose you come up with an idea for a new business. To whom would you turn? You would probably turn to people with whom you already have a personal relationship. You would try to recruit them to join your organization and encourage them to help recruit competent people they know to the organization. Once the formal organization is established, it in turn gives rise to informal communication among strangers, which establishes the informal organization.

Barnard also taught that a healthy informal organization is necessary to effective formal organization. The first indispensable function, said Barnard, is communication: the formal channels of communication can't begin to handle all the coordination that needs to be done. Informal communication complements formal communication. The second function is to maintain solidarity, high morale, and a willingness to serve. A third function is to protect the "integrity of the individual" (Barnard 1938, 122).

[handwritten margin note: functions of informal com]

Organizations often order people to do things that violate the values and integrity of the individual. An informal network of communication can give an individual the support she or he needs to resist the formal system. Informal groups sometimes set production quotas of their own that are less demanding—and safer—than those established by management.

I saw some unusual informal organizations at MSFC. In the most interesting case, American managers complained to me that the German-born managers had a tight informal system because of their native language, customs, and long history together. The Germans socialized on a regular basis, and the Americans felt "the family," as they called them, sometimes did organizational business at such events. The organization was open enough, however, for me to bring this topic to the attention of the entire management group, Americans and Germans alike, and this stimulated a healthy discussion of the problem. The other observation I made was that the engineering community of the organization supported each other via an informal communication system; this gave them the confidence to raise red flags about technical problems the formal system hadn't discovered.

My initial meeting with von Braun was scheduled for July 24. I had been curious about his involvement with the Nazis before and during World War II in Germany. Wiesman said that although von Braun had to join the Party, he wasn't a true believer. He told me that von Braun had been arrested by Heinrich Himmler's SS for "defeatism," for making a suggestion at a party that the V-2 Rocket they were working on could be used for space travel after the war. General Dornberger, the military leader of the rocket program secured his release, and von Braun's friends and defenders said this proved his innocence and distance from the Nazis. Later I would learn that the situation was more complex than that. By 1943 the lack of labor for producing the V-2 missiles was a problem. Arthur Rudolph, chief engineer at the Peenemünde factory, learned of the availability of concentration camp prisoners, and "enthusiastically endorsed their use, and helped win approval for their transfer" (Dunar and Waring 1999, 7). They were taken to the central V-2 plants, called Mittelwerk, in the Harz Mountains. Atrocities that occurred there plagued the reputation of the German team for the rest of their days. Von Braun admitted that he visited the plants, but never saw any mistreatment, much less any atrocities. As two historians of the Marshall Center wrote:

> Von Braun's relationship to the Nazi Party is complex; although he was not an ardent Nazi, he did hold rank as an SS officer. His relationship to slave labor is likewise complicated, for his distance from direct responsibility for the use of slave labor must be balanced by the fact that he was aware of its use and the conditions under which prisoners labored. (Dunar and Waring 1999, 7)

When the time arrived, Walt Wiesman had arranged the appointment and walked me over to the tall, gray cement-and-glass building housing the management group. We ran into von Braun outside his office on the top floor. He was a large handsome man with a large smile. I noticed a scar above his lip; I later learned that he had been injured in an automobile accident during the race to avoid the advance of the Red Army of the Soviet Union on Peenemünde; von Braun and his top people had had enough of totalitarianism, so they sped toward Bavaria to surrender to the American Army.

He invited me into his office. As Wiesman started to leave von Braun called out for him to stay and listen to the interview. "Don't be so goddam lazy," he joked. Von Braun sat at the head of the table; his assistant Jim Shepherd and I sat on the left side, Wiesman opposite us on the right. To my opening question about the importance of organizational communication von Braun responded in the following way:

> In such a dynamic organization you have to keep up, keep the organization informed from top to bottom. People with problems and suggestions must be able to get the attention of top management. Communication up must be free, not tied to channels, if management is to be kept informed. However, there must be clear action and command channels. Closed and open loops. For

interesting

example, there are weekly reports from lab directors and project engineers in RDO and from project engineers in IO which bypass layers of management, but directives go back down through Mr. Weidner [Director of RDO] and General O'Connor [Director of IO].

It would take weeks for me to understand completely what von Braun meant in practical terms by "bypassing," "action and command channel," and "open and closed loops." He spoke rapidly with a slight accent, praising Walt Wiesman's five-year organizational communication plan. It was clear to me that von Braun considered upward-directed communication to be crucial. In fact, I would later hear from Marshall managers a story that is no doubt fictional but needs to be retold because it is true to the organization and its communication practices.

'*Five Nines.*' The story was related to me during an interview later that summer. The interviewee said it was probably apocryphal but deserved to be true because it so perfectly captured the way communication worked at the Marshall Center. It seems von Braun was attending a meeting at NASA headquarters in Washington, D.C.; someone asked him for the reliability figure for a particular rocket stage.

"I don't know," von Braun answered, "but when I get back to Marshall I'll find out and call you."

After he returned to Huntsville, he informally called several of his long-term colleagues—some were several levels below him in the hierarchy. Then he phoned Washington. The NASA official heard von Braun's answer as "five nines," shorthand he assumed for a reliability figure of 0.99999.

"Fine," said the official. "How did you arrive at that figure?"

"Well," answered von Braun, "I called Walter Haeusserman in the Astrionics Lab and asked him, 'Are we going to have any problems with this stage?' He answered, 'Nein.' Then I posed the same question to Karl Heimburg in the Test Lab and he also said 'Nein.' I kept at it until I got five Neins." Although I'm sure it didn't happen that way, people told me the story because it captured the climate of the management system. It also illustrates vividly those three close cousins in the Ideal Managerial Climate: Trust, Confidence, Credibility.

Upward Communication as an Earthquake Prediction System

Von Braun developed the point, characteristically, by means of an analogy:

> This is like being in the earthquake prediction business. You put out your sensors. You want them to be sensitive enough, but you don't want to get drowned in noise. We have enough sensors, even in industry. There are a lot of inputs about trouble. Some are too sensitive; they overact. Someone else might underestimate. You want to know the name of the guy. Is he one of the perennial panic-makers? Some guys always cry for help. You need balance in the system—to react to the critical things. Exposure teaches you how to react. Some create problems and then proudly announce they have solved them. Others make a lot of noise just to get the mule's attention.

As I typed up my copious notes after the interview I thought to myself, *he must have heard the story about the farmer who hits the mule in the head with a two-by-four to get his attention.* I also realized that he had a keen understanding of source credibility, an important communication factor, in his appreciation of the need and ability to read his "sensors," or sources, carefully. Notice the analogy to sensors, the devices that gave the shuttle managers the first indication of trouble with *Columbia.*

Von Braun urged me to interview key people in the organization in the interest of finding blockages, barriers, or other problems in organizational communication. I was warned, however, that a recent questionnaire survey of morale had irritated the engineers and scientists. They didn't like the simplistic forced-choice answers. It was decided I should interview the top 50 engineers, managers, and scientists at Marshall. My interview guide would be fairly open-ended, asking, "What is working well and what isn't?" Answers to these questions would be followed up with probes. If a particular topic didn't come up, I'd ask a direct question in neutral language. The interviews would be anonymous in the sense that I wouldn't reveal who said what. Von Braun sent a letter to each member of the sample summarizing my credentials and asking each person to cooperate. They realized I would synthesize the results and report back to von Braun and eventually to the entire top Center management. That encouraged them to open up to me, knowing as they did that their problems would have the mule's attention.

Shepherd's Three Problems

Before I started the formal interviews, Jim Shepherd, von Braun's assistant, who had sat in on my initial interview with von Braun, called me over to his office. He wanted me to help him solve three particular problems while working concurrently on the larger study.

The first problem had to do with the Saturn V Control Center. I had been given the standard tour of the facilities during my orientation and was highly impressed. It was a huge room with an oversized conference table in the center. The walls were covered with floor-to-ceiling charts. These graphic displays tracked every vehicle of the Saturn V rocket series as well as the time intervals between events, using PERT (Program Evaluation and Review Technique), a technique widely used by the Defense Department, NASA, and industry. A crew of attendants posted milestone stickers on the charts as each step was completed. The charts indicated what tasks had moved into the "critical path" and who was responsible for completing them. These displays also showed the remaining steps and their completion deadlines in order to meet the schedule for the project.

Shepherd explained the problem. The Administrator of NASA, James E. Webb, was trained in public administration. He believed his contribution was to help create new methods of management. He and others believed that the management system of MSFC was informal, even primitive. "Administrator Webb believes in the Control Center so much that he brought 75 university people to see it. He wants them to create a curriculum around it and have students come in and critique us. But we aren't all that sold on the Control Center. Mr. Webb wants to establish a total management concept for government. So the first question for you, Phil, is how valid and timely are the data in that room?"

I said I'd work on it.

The second problem was about briefings and reviews. "You get a briefing and it's canned," Shepherd complained. "The briefers use alphabet soup: acronyms like ICN and ICD and so on. I don't believe half the guys hearing a briefing understand it. Even von Braun doesn't understand some of this stuff. We need a set of guidelines for briefings."

Again, I said I'd work on it.

The third problem had been created by the success of what were known as the "Monday Notes," a reporting system explained in detail later in this chapter. After the elaborate investigation of the fire in which astronauts Grissom, White, and Chaffee were killed, everyone came to the realization that von Braun was the best informed person in NASA. People asked how that was possible, and they found out about the Monday Notes. As a result, Administrator Webb decided that NASA headquarters should receive a similar weekly report from all field centers. "We're supposed to send it to Mueller," Shepherd said, "and he'll excerpt it for Webb. But the problem is what should we send them? What should we take from the notes to von Braun? If the material is misunderstood or taken out of context, what problems will that cause for us the next week?"

I again agreed to work on it after getting Shepherd to explain who Mueller was. He was George Mueller (pronounced "Miller"), of NASA headquarters, who became Associate Administrator for Manned Space Flight in September 1963. His training was in electrical engineering. Von Braun reported to Mueller on all matters related to manned flight.

I was glad that Shepherd had given me these problems to work on. Now I would have some specific, directive questions to ask if my general questions drew blank stares. It also meant that I would have to get back to Shepherd with a preliminary report before making my more general report to von Braun.

As I began the interviews, moving from office to office and from lab to lab, I began to understand the dynamics of the organizational communication system. I conducted interviews with top management, the directors of RDO and IO, the director of each laboratory and RDO office, the chiefs and managers of the IO offices (with the exception of Arthur Rudolph, the Saturn V Program Manager), and the chiefs of all the staff offices. The reader will note the preponderance of German surnames in the following *partial* list of interviewees. This conveys a flavor of the culture and structure of the organization in 1967.

Director: Wernher von Braun

Deputy Director, Technical: Eberhard F. M. Rees

Deputy Director, Administrative: Harry H. Gorman

Director, Research and Development Operations (RDO):
H. K. Weidner

Director, Industrial Operations: (Gen.) E. F. O'Connor,
Air Force

Director, Aero-Astrodynamics Laboratory: E. D. Geissler

Director, Astrionics Laboratory: W. Haeussermann

Director, Computation Laboratory: H. Hoelzer

Director, Manufacturing Engineering Laboratory:
W. R. Kuers

Director, Propulsion and Vehicle Engineering Laboratory:
W. R. Lucas

Manager, Mission Operations Office: F. A. Speer

Manager, Michoud Assembly Facility: G. N. Constan

Manager, Mississippi Test Facility: J. M. Balch

The last two interviews necessitated trips to New Orleans and Hancock County, Mississippi. I opened each interview with this question: What works well? And followed up with: What isn't working well? Then I would become more directive, narrowing my questions to Shepherd's three problems and other items mentioned by an interviewee that needed to be probed.

The Mushroom Problem

I showed up for one of my interviews with a lab director early one summer morning and was shown into his office.

"Ah, Dr. Tompkins," he said in a German accent. "I understand you are here to see me about the mushroom problem."

"There must be some misunderstanding," I replied. "I don't know anything about a mushroom problem."

"Aren't you interested in organizational communication?"

"Yes."

"Do you know how to grow mushrooms?"

"No," I said, wondering what was happening to my interview.

"The way mushrooms are grown illustrates a common problem of organizational communication. You put them down in the basement and keep them completely in the dark. Every once in a while you open the door and throw some horseshit on them."

We both laughed heartily and had a productive interview, talking about the mushroom problem and how to solve it.

The Monday Notes

The answer I got in my interviews to the question of "What works well?" invariably was the "Monday Notes." The Monday Notes came into existence during the growth stage of the Marshall Center. Kurt Debus, one of the "Peenemünde Gang," as von Braun called them, had always been in charge of launch operations for the organization. During a period in which Debus was away at Cape Canaveral supervising a launch, von Braun felt that the absence of propinquity diminished vital communication between the two of them. He asked Debus to send him a weekly one-page note from the Cape, in which he would describe problems and progress of the past seven days.

Finding that he looked forward to reading the weekly note, von Braun decided that a similar note from other key personnel would help him keep informed and get important issues out in the open. He asked about two dozen managers (all of whom I interviewed), lab directors in RDO and project managers in IO who were removed from him by at least one layer of management, to send him a weekly, one-page note summarizing the problems encountered and progress made during the week. Simplicity was the key. There was no form to be filled out. Requirements: no more than one page, headed by the date and the name of the contributor. They were due in von Braun's office each Monday morning. In this way a layer of management was bypassed on the way up. (The Directors of IO and RDO were by-passed but could read the notes; they just couldn't change them in any way.)

As von Braun read each note, he initialed it with a "B" in the right-hand top corner and wrote the date in numbers. (The first time I saw a set of the notes, the teacher in me automatically flipped through them looking for an "A" paper.) He also added a considerable amount of marginalia in his impeccably clear handwriting, making suggestions and dishing out praise. The notes of July 10,

1967, for example, include a hand-written question directed to a manager about vehicle cost figures included in his note: "Have we passed this on to Mueller [a NASA headquarters official]? B." There is the suggestion to another manager that a new computer mentioned in his note "could be immensely useful for earth resources surveys from orbit. B." To a lab director whose earlier recommendations about the superiority of one type of weld over another had been rejected and who had conducted additional tests supporting his original position, von Braun wrote, "Looks like you won after all! Congrats. B."

All 24 notes were annotated and collected in alphabetical order by the authors' surnames; they were then reproduced and returned as a package to all of the contributors. What was the organizational effect of this simple, innovative activity? I put that question to all of the contributors as I interviewed them. The answer was an almost unqualified praise for the notes. The reasons offered are worth considering.

The first virtue is obvious. The notes served as one more channel—24 lines of communication—in addition to briefings and memoranda, to keep the boss informed of problems and progress. Whereas many organizations suppress bad news, this one sought it out. It thus fostered another factor in the Ideal Managerial Climate—Openness and Candor. Another benefit pointed out was the crucial lateral or horizontal function of communication. Each lab director learned what all the laboratories had been up to during the previous week; in addition, units in IO could read about the activities in RDO, and vice versa, which often stimulated additional horizontal interaction among them to maintain coordination of projects. A good illustration of this occurred during the interview process. A lab director I was interviewing said the recently returned Monday Notes contained a comment from von Braun suggesting that he needed to telephone a project manager in IO about a mutual problem. By chance, my next interview happened to be with that very project manager. Without prompting from me he volunteered that he discovered a suggestion in the notes that he needed to telephone the lab director whose office I had left minutes earlier.

The marginalia supplied by the charismatic von Braun made the notes "the most diligently read document" at MSFC, as a lab director said to me in an interview. This crucial *feedback function* was mentioned by nearly every contributor. They saw how the boss reacted to the week's work in their own units as well as his response to other contributors' notes. Another desirable effect mentioned by most of the interviewees was closely related to the feedback function. Because von Braun found it increasingly necessary to travel to California, the Cape, Washington, D.C., and other locations, the Monday Notes "kept the channels open" during the periods of limited face-to-face interaction.

The notes also were an antidote to the sterile, formalized procedures that permeated government work. Contributors derived a high degree of communication satisfaction from the *personalized* nature of the notes, a break from the code-numbered, U.S. Government-NASA-MSFC format designating the lab or office from which it emanated. Just the name and date at the top of the page. The contributors also derived what Redding called communication satisfaction from the *informality, quickness,* and *frankness* of the notes. In regard to frankness, some rather fierce arguments were carried out in the notes. One unit's note of the current week might challenge another's note of the previous week. This controversy, said a lab director, gave the notes their particular "charm." Another lab director said with a disingenuous smile, "We sometimes misuse them—to get attention." *The two-by-four and the mule again,* I thought. In other words, the notes provided a *public forum* and a *court of last resort.*

During my interviews I discovered something von Braun and most of the others didn't realize. Curious about how the 24 contributors generated the content of their weekly notes, I systematically asked about their procedures. In most cases the lab director would ask his subordinates, the division chiefs, to provide him with a Friday Note about their activities. Von Braun was delighted to learn of this unexpected dividend when I briefed him later in the summer. (During the summer of 1968, I had the opportunity to interview 14 division chiefs. Most of them requested a similar note from their subordinates, branch chiefs, and so on.)

Moreover, some of the directors and managers organized meetings to determine what should be put in next week's note and to discuss von Braun's responses to the most recent packet of notes. Relevant portions of the notes were reproduced for distribution down the line. In short, von Braun's simple request for a weekly note had generated a *rigorous and regularly recurring discipline of communication within the organization.* Once a week, almost every supervisor in the organization paused to reflect on what needed to be communicated up the line. Lab directors and project managers stopped to read what their peers had communicated to the director and how he responded.

The Monday Notes also illustrate two general principles in von Braun's philosophy of organizational communication—*conflict* and *redundancy*—that were manifested in other ways as well. As indicated above, some rather heated conflicts surfaced in the notes from time to time, as well as in briefings and memoranda. My interviewees sometimes tried to draw me into the conflicts in hopes that I would advocate their side to von Braun. Occasionally I did. Usually I intervened when von Braun asked me to look into this or that controversy. Conflict was natural and necessary to such a large organization, and some of it was petty. Von Braun had a positive attitude toward conflict and sought to encourage it; perhaps he had even engineered the means to impress it into the very structure of the organization.

Monday Notes as Organizational Learning

In a book chapter about organizational learning, Karl Weick and Susan J. Ashford (2001) used my published descriptions of the Monday Notes (1993) to illustrate their theory; in their summary they paused to write, "Tompkins then goes on to report an intriguing discovery" (725). The discovery was the new understanding of how the contributors developed the content of their notes, and the resulting discipline of communication. Weick and Ashford make this interpretation:

> Since the notes were written weekly, they are closely attuned to the flow of events that occur in a relatively short period and, therefore, accurately represent continuity. Since drafting the notes is done in part by people who are actually doing the work to be represented, the drafting is constitutive as well as representative. People do things they can talk about. . . . In the Monday Notes, the flow of events through Marshall is mapped accurately by communication practices that mirror them. The result should be faster learning. . . . Because the notes spanned units with quite different missions and know-how, their communication languages also differed. Therefore, when people read the notes they are reminded of several different ways to frame common problems. This guards against some of the traps mentioned earlier where groups get stuck in a narrow language, such as the language of profits, and are unable to see any other meanings that could help dissolve or

solve their problems. The Monday Notes also potentially enlarge the domain of what is *discussible* (emphasis added), and bring strategies and assumptions into play, as well as tactics. Thus, learning can engage more fundamental understandings. (726)

Weick and Ashford then introduced a negative concept, a serious hindrance to organizational learning that they think could be overcome by practices enacted in the Monday Notes. The idea is called "hindsight bias."

> Hindsight bias, which involves the use of knowledge about outcomes to edit reconstructions of the antecedents of those outcomes, should lead people to learn the wrong things. . . . Hindsight bias should be reduced by the practice of Monday Notes. . . . In hindsight, there appears to be one best way and nothing much to learn. This conclusion is troublesome because it was arrived at through severe editing out of complexity and ambiguity present at the time the action originally unfolded. Those complexities might suggest the wisdom of different choices in the future. Unfortunately those complexities can't be retrieved once justification has *masked* them (emphasis added). However, frequent communicating, such as is represented by the Monday Notes, is one way to prevent such masking. Neither much history nor much justification are allowed to build up around a choice before it is subjected to weekly public scrutiny, criticism, praise, and alternative constructions by von Braun and his associates. It is easier to keep perceptions and actions separated when progress is reported frequently. If they are kept separate, this should make it easier to experiment with new perception-action linkages and improve performance. (726–27)

Weick and Ashford thus show the connection between communication and learning. The climate and culture of an organization, developed by the way in which people interact, can determine how well the members learn what is important to achieve high-performance goals. Crucial to this at MSFC was the candor, openness, and a healthy attitude toward conflict. These factors of the climate or culture could be seen and heard in oral discourse as well as in writing. An example would be the briefings I attended at the Marshall Center in 1967 and 1968. Briefings became a site of a verbal agon, or struggle, between briefer and those being briefed and among members of the audience. I remember in particular a case in which a senior briefer from one laboratory was questioned by a young engineer in the audience from a different unit. Von Braun was present during the exchange, turned to the young man, restated what he thought was the point the young engineer was making, and after getting positive feedback he turned toward the briefer and told him he and his colleagues should give attention to what the young man had said.

I was asked by von Braun to give a two-hour briefing to the top 50 people at the space center after finishing my research. I was interrupted time and again with questions and objections in a lively give and take. As important as the Monday Notes were, there were other practices that contributed to the culture of open communication.

Automatic Responsibility

One of the most remarkable concepts in von Braun's philosophy and practices of organizational communication came up during my interviews. In addition I had come across a paper von Braun had given at a conference at French Lick, Indiana, entitled "Building a Research Team." In it von Braun said that at the Marshall Center, the laboratories

> have full cognizance and *responsibility* for all efforts that fall within the purview of their respective disciplines, including active projects, further project studies, and supporting research work. . . . [The lab director] is expected *automatically* to participate in all projects that involve his discipline and to carry his work through to its conclusion. (emphasis added)

My interviewees all referred to this concept in interviews as "automatic responsibility," explaining that it had long been part of the organization's method of operation. In practice it meant, for instance, that an electrical engineer working in the Astrionics Lab assumed responsibility automatically for any problem he perceived to fall within his area of competence, regardless of whether the task had been assigned to his lab. He was expected to live with the problem until it was solved. I heard a story about the case of a young American-born engineer who assumed responsibility for a problem and wound up recommending a solution opposed both by his superiors and much of the technical talent at the Center. In the end, however, he was found to be correct and was subsequently rewarded by the German leadership group with a top management post.

If the person who perceived a problem lacked the technical ability to see it through to its conclusion, he or she then assumed the responsibility for communicating his or her perceptions of the problem up the line so that top management, thus alerted, could direct the appropriate specialists to it. Automatic responsibility sometimes ruffled the feathers of certain managers, who felt that other people should tend to their own business, but such irritations were worth the price because the system guaranteed that problems would receive the attention they deserved.

This practice was such a radical innovation that I've found few executives who felt they could live with such a principle. It sounds too much like anarchy. Better to stay with more bureaucratic practices. In addition, the fact that it worked at the Marshall Center proves that able people don't necessarily avoid responsibility. To the contrary, it proves that people will in the proper culture actively and aggressively seek responsibility for the good of the organization. We now have a tentative answer to the question raised in Chapter Three, the question posed by Eisenberg and Riley: Can a "strong" culture be flexible and sensitive enough to adapt to a changing environment? My answer is *yes, if the strong culture is also equipped with communicative practices such as the Monday Notes and Automatic Responsibility, practices that make it a learning organization.*

Penetration

Part of the effectiveness of the Marshall Center was due to a concept called *penetration.* I had found the practice documented by earlier students of NASA, writers who discovered why the Marshall Center's private contractors performed better than they did for other centers and other government agencies, including the Air Force. A book by H. L. Nieburg, *In the Name of Science* (1966), provides among other things a catalogue of atrocities committed by the Contract State. The book reviews systematic waste, profiteering, technical failures, cost overruns, and plain ineptitude in the work of corporate contractors, mainly in the defense industry and the government agencies that were supposed to be monitoring them. Nieburg narrates a congressional subcommittee inquiry into how a specific contractor, General Dynamics Astronautics, could have made some fundamental mistakes in its development of the Centaur rocket for the Air Force. The project was transferred to NASA and von Braun was called to testify.

According to Nieburg, von Braun's testimony "euphemized the situation yet succeeded in conveying a sharp indictment of the Air Force management" (275). Von Braun's words were:

> I think what we felt was a lack of depth of *penetration* of the program on the part of government personnel, in very general terms. We believed this staff of eight people [was] just an inadequate coverage on the part of the government, no matter whether it is NASA or the Air Force. (275, emphasis added)

This is an excellent example of how different MSFC's methods of management-communication were from the ones used by the Air Force. When NASA assumed the management of the project, von Braun assigned 140 technical employees to supervise and penetrate the contractor (in contrast to the Air Force's eight).

An even more dramatic illustration of the practice of penetration emerged in one of my interviews in 1967. A top-ranking manager explained to me that a contractor's first experience with the Marshall Center was usually "traumatic" because the Marshall engineers were so aggressive in assuring that they were going to get exactly what the government ordered. One such contractor delivered a rocket stage to Huntsville. This stage had rigorous "heat leak" specifications that virtually required a vacuum seal; the requirements were so exacting that an ice cube placed inside it would take eight and a half years to melt and another four years for the water temperature of the resulting puddle to rise to 70 degrees. The Marshall personnel quizzed the contractor on the possibility of cracks in the stage. The contractor's people finally admitted there might be some cracks in it. How many cracks?

"Twenty-one," was the answer.

"No, there are 26 cracks," asserted a Marshall engineer.

The rocket stage was submitted to an X-ray examination. The stage contained 26 cracks—along with workers' tools and lunch boxes and other debris that shouldn't have been there. How could the customer know more than the manufacturer about the product as it was delivered? The answer is that the customer had penetrated the manufacturing organization, observed the production process in the contractor's plant, monitored workbenches, and quizzed the contractor's employees who were willing to talk about their problems. The contractor's management, by the way, was reportedly well known for discouraging the upward flow of bad news. Its top managers had apparently been forced out after the Apollo 204 fire for killing messengers who brought bad news.

There were *intra*organizational applications as well. A thoughtful lab director who had read widely in the management literature explained his personal adaptation of the principle. Just to get his job done he had to penetrate two layers up and two layers down. He qualified the principle by explaining that in penetrating two layers down, to his subordinates' subordinates, his purpose was not to direct or command them. Instead, he listened and talked in order to keep up with them. Another lab director told me that his group was falling behind the pace of the Center because he hadn't been given sufficient lead time to erect a large test stand for the RDO people. The lab director solved this problem by means of intraorganizational penetration. He persuaded a German engineer to come to work for him and then dispatched the man to attend all reviews and briefings and keep him informed so that he could anticipate future needs. This practice of penetration enabled the unit to keep ahead of the rest of the organization.

A Briefing for Shepherd

I met with Jim Shepherd on September 5, 1967, in his office adjacent to von Braun's and briefed him on the three issues. We started with the Saturn V Control Center, and I gave him an outline to use during my briefing. My evidence came from interviewees' perceptions of the Control Center plus my own experience during most of an afternoon spent in the Control Center, interviewing other middle- and upper-level personnel who worked there. I had found the personnel to be remarkably candid about the limitations of the center. After explaining my methods to Shepherd, I opened the argument with my conclusion: "The Saturn V Control Center is more for show than utility."

The Saturn V Control Center Problem

PERT (Program Evaluation and Review Technique) had been abandoned, I explained, because Arthur Rudolph (I never met the man who had endorsed the use of slave labor—he was always out of town when I tried to see him) said he didn't understand it. The staff had replaced the PERT chart on the wall with cartoon cutouts and a system called the "waterfall" that Rudolph had used in Germany. I was told he visited the Control Center about three times a week. I asked Rudolph's assistants where Rudolph normally spent his working day. They led me across the hall to a room about a tenth of the size of the Control Center. In contrast to the empty Control Center, the room was alive with activity: people were on the telephone, talking to contractors and NASA personnel all over the country, making notes and then translating them onto the grease-pencil charts that almost filled the room. Here I had found the beehive, the real control center for Saturn V. Rudolph began and ended the day in this informal mini-control center, and I was told he kept another chart in his own office.

I informed Shepherd that the information on the walls of the official Control Center was outdated by as much as three months. The reason for this was that in some cases information had to flow from the contractor to the resident civil service (NASA) manager to the Control Center and then to the people who plugged the data into the PERT chart. PERT then also had to be translated into Rudolph's "waterfall" system. Staff members admitted Rudolph frequently discovered discrepancies between what he knew and what was visibly displayed in the Control Center. For the up-to-the-minute status of a problem, one had to call as many as three Saturn V stage managers.

Further, the information in the Control Center was, as I explained to Shepherd, in some cases invalid. If someone in a critical path didn't meet a deadline or "milestone," it was called a "slip" in the schedule. The staff in the Control Center wouldn't be likely to admit a slip in the schedule because they wouldn't want the subsequent units in the critical path to relax. Rather than acknowledge a slip of a week's time, they would parcel it out day by day. Consequently, opportunities for improving quality and reliability were lost. (In one of my interviews, a lab director was concerned about a swing arm on the tower that held the Saturn V in place until the actual launch. He knew it could be rebuilt and improved in just a few days but couldn't obtain permission to do it because the slip was being meted out day by day, one day after another.)

Milestones weren't posted, I continued, until all the formalistic documentation had been completed, even though the work was finished and had been adequately evaluated. A lot of paperwork was required to certify the completion of a task, and postings of milestones reached were delayed until the necessary paperwork reached the Control Center, which took time. Hence, work that was done might not be formally displayed as such. In addition, everyone I talked to admitted that contractors played games with their inputs, i.e., they manipulated time frames by denying they were behind when in fact they were.

I presented Shepherd with evidence from the formal interviews that Marshall managers usually knew the information on display was inaccurate when they went to the Control Center for a briefing. But their most acrimonious complaint was that the Center's intimidating atmosphere or ambiance prevented open discussion. "You can't argue or compromise with PERT or waterfall," they told me repeatedly. The result was a classic case of goal-displacement. Methods that were devised to serve as management tools, as *means* to organizational *ends*, had become ends in themselves. There was grave concern among the interviewees that *things* rather than *people* were threatening to gain control of decisions and events.

Shepherd was delighted with the exposé. He was also convinced by the argument and the evidence. He went on to explain that Webb's background in public administration motivated him to claim advances in management; he was fond of bringing members of Congress and other VIPs to Huntsville to show them the Saturn V Control Center as NASA's contribution to new management-communication developments. Only recently did I discover that some historians of space claim today that the "secret" of Apollo was systems management and other formalisms (e.g., Johnson, 2002). The moon rocket, however, was managed informally in a small room filled with people handling telephones and grease-pencil charts.

Briefings and Weekly Notes for Headquarters

I dealt with the briefing problem by giving Shepherd a set of guidelines for such presentations. Readers interested in pursuing them may find them in *Organizational Communication Imperatives: Lessons of the Space Program* (1993). I told Shepherd I couldn't offer much help about how to prepare the weekly notes for Washington. The contributors to the Monday Notes were alarmed by the possibility that their notes might be excerpted and relayed to headquarters out of context. Some said they would consider putting an asterisk or bullet next to items that might be forwarded, but their concern was that Washington might well misinterpret some of their "behind the scenes" intra-center quarrels and statements of alarm—the very qualities that gave the Monday Notes their "local charm" and much of their usefulness.

Shepherd told me that since giving me the assignment regarding the preparation of weekly notes for Washington, NASA headquarters had lost interest, and the plan had been abandoned for a weekly note from each of the field centers. Headquarters did, however, direct the other field centers to emulate MSFC by establishing their own set of weekly notes. True to formal, bureaucratic form, they produced a full volume of directions, including numbered forms, about how to initiate them. The bureaucrats seemed to miss the point of the advantages of informality and simplicity. Shepherd was appreciative of the briefing and added that I'd already earned my GS-13 salary for the summer. He planned to sit in on my briefing for von Braun later that week, and looked forward to learning what else I had discovered. He urged me to emphasize all the problems I had uncovered in my interviews to von Braun.

Briefing for von Braun

The briefing was scheduled for 3:30 p.m. on September 8, 1967, in von Braun's office. I arrived early to put my entire outline on the blackboard—consistent with the briefing guidelines I'd recently given Shepherd. After discussing the methods and limitations of the study, I discussed "What We Do Well." (My heading reveals that I had already slipped into the use of the "we" of identification.) Then I moved on to "Problems and Challenges." The first problem I discussed under that heading was "The Formalism-Impersonality Syndrome." I told von Braun and other top managers he had invited that it was refreshing to hear people who had made important contributions to the development of computers and management-information technology sternly denounce the sterility, formalism and impersonality of many of the management tools they used, some of which had been imposed on them by NASA headquarters. Then I moved on to the specifics: overuse of computers, control centers, and "canned" reviews. I made the same presentation I'd given Shepherd about the Saturn V Control Center. Shepherd smiled. Now I had their attention.

Von Braun immediately challenged me about the Control Center. Using an analogy to newspapers, he pointed out that a certain degree of obsolescence was necessary. I replied, "But that's my point. If it's obsolescent, it must be more for show than utility." We argued back and forth, and several others in the room began to take my side against von Braun. Later I learned that von Braun had been convinced by my arguments early on; this dialectical devil's advocate approach was his way of testing the strength of my evidence and arguments. (I also learned later that von Braun stopped attending briefings at the Saturn V Control Center. It became Administrator Webb's White Elephant, a place where he brought VIPs for show-and-tell sessions to keep them out of the way of the busy teams getting ready for the moon launch.)

As I started to move on to the next problem, von Braun interrupted again, asking, "Wait, isn't there something we can do about this?" I reminded him of my approach by pointing to the outline on the board, saying that I wanted to present all ten problems and then discuss ten recommendations. (The reader who wishes to learn of the other nine problems and my ten recommendations can find them in *Organizational Communication Imperatives*.)

The briefing took two hours because of the back and forth debate about my claims. As it came to an end, von Braun shook my hand, congratulated me and asked when I would be going back to Detroit. When I answered that I'd be leaving in a few days, he said that he hoped I would be willing to come back to make the same presentation to his next Staff and Board Meeting—the top 50 or so managers at MSFC—on October 6, 1967. I agreed to do so. He made it clear that he didn't want to edit or censor my presentation in any way, although NASA headquarters wouldn't be happy to hear about one of the problems I discussed—it dealt with the German informal organization under the heading of "It's a Family Affair." (This was one of the first indications I'd seen about how sensitive this issue was in Washington.) He did want to talk about the order in which I would present the problems and recommendations to the Staff and Board meeting. The others who heard the briefing had ideas about the order as well. In the end, we agreed that because of its importance, the problems of lateral or horizontal communication would be the first presented.

When I left for Detroit I was both proud and apprehensive about being asked to return in three weeks to brief some of the world's top rocket scientists about their organization! I also felt a strong sense of identification with MSFC, NASA, and Apollo. Von Braun sent a gracious letter to the president of Wayne State University praising my work and asking him to allow me to return for another briefing.

Staff and Board Briefing

As I rode the elevator to the ninth floor of the management building on October 6, 1967, my apprehension mounted. In a few minutes I would be addressing 50 or 60 of the brightest rocket experts in the country. Heightening my anxiety was the fact that some of my remarks would be critical of some very powerful individuals in the room.

Von Braun played on that theme as he introduced me. After mentioning my credentials, he explained that he had received a briefing from me three weeks earlier and had asked me to present it to the Staff and Board. He described my report as so "hard hitting" that he felt it necessary to ask that no one leave the room. The tension in the room grew tighter. Then von Braun deadpanned that because an important phone call was coming in from NASA headquarters, he would have to leave the room for a few minutes. The audience exploded with laughter, releasing some of the tension.

There were three huge screens behind me. Following the briefing guidelines I had given Shepherd, I put my overall outline on the screen on the left. I used the other two screens, at the appropriate moment, to display the enumerated problems, then the recommendations. I started with what worked well, limiting myself to a discussion of the Monday Notes. As I finished, I was interrupted by a sharply spoken question voiced in a German accent: "Is that the only thing we do well?" I realized then that we were playing hardball.

"No," I snapped back, "you do other things well too. But I don't want you to feel too good about yourselves at this point." In the raucous laughter that followed, I began to relax. We marched through the problems one by one. As I made my case, some participants raised objections while others came to my defense, supplying evidence to support my analysis, all in a medley of German and American regional accents. Soon I recognized the unmistakably high-pitched voice of von Braun, who had slipped back into the briefing room without my having noticed. It was a good give-and-take, typical of an MSFC briefing. (The entire two-hour session was audiotaped by MSFC, allowing me to reconstruct the dialogue in *Organizational Communication Imperatives.*)

Saturn V Flies

On November 9, 1967, I got up early to watch the first unmanned launch of the moon rocket, the Saturn V, on television. It was a tense situation because of "all-up" gamble in this flight imposed by George Mueller of NASA headquarters. With previous multi-stage rockets, each stage was successfully tested in flight with "dummy" stages above it. This would, of course, take additional time. With the first flight of Saturn V, all three stages were being tested during the same launch. If everything went well, NASA would be on schedule to keep Kennedy's promise to land a man on the moon, and return him to Earth, within the decade. If not, the program was in trouble. I sat down in front of the television set and listened to Walter Cronkite count down the launch. Even the speaker in my tiny television set conveyed the thunderous thrust of the five F-1 engines firing simultaneously, so powerful that Cronkite announced that vibrations were bringing down the ceiling of his control booth. The Saturn V soared into the heavens, each stage functioning as designed. We were on schedule.

Return to MSFC

I agreed to return to the space center in Huntsville in the summer of 1968. Many of my recommendations made the year before had been implemented. This time they were ready for me. Dr. von Braun had a set of research questions and what he called "action items," such as should the Computation Laboratory be allowed to design hardware for boosters and spacecraft as well as the Astrionics Lab? (After my research I said "yes.") Von Braun also handed me a document entitled "Questions Related to the Organization and Implementation of the Systems Engineering Effort and Related AAP [Apollo Applications Program] Management Areas at MSFC." The three-page paper consisted of what became known as "The 24 Questions." Now that the work for the moon rocket was nearly completed, MSFC had to reorganize for the new Apollo Applications Programs, a three-component cluster called Skylab, the Orbital Telescope Mount, and a docking adaptor. Washington had made it clear that these AAP components would have to be developed with fewer dollars and fewer people. George Mueller, von Braun's boss in NASA headquarter, wanted MSFC to strengthen a discipline called Systems Engineering.

The reorganization would take into consideration the downsizing and the need to give systems engineering greater authority. There was a feeling that the laboratories sometimes optimized their components in such a way as to make the interfaces or connections between them difficult to match. Von Braun had always served as the head systems engineer of the team, thinking through the ways in which subsystems would have to be designed in such a way that they would "fit" into the overall system. He gave me some other jobs to do, but this year he wanted me to answer the 24 questions about reorganization, concentrating on the place of systems engineering.

So it was back to work. I conducted another 50 in-depth interviews, attempting to discover a new scheme of organizational communication that would give systems engineering a central role in future R&D work. Perhaps the best way to describe my work in 1968 and my three-hour briefing for von Braun is to summarize and quote what a space historian had to say about my efforts. His heading was: "Von Braun's Conversion."

> By the summer of 1968, von Braun recognized that he needed to strengthen systems engineering at MSFC. He called in Philip [sic] Tompkins, a communications expert from Wayne State University, to study MSFC's organization and recommend how better to implement systems engineering. At the time, Mueller was pressing von Braun to emphasize systems engineering in the design of the *Skylab* space station. Von Braun explained to Tompkins that Mueller, who had been trained in electrical engineering, thought more naturally in terms of a "nervous system" than he, who thought of rockets as machines. Von Braun belatedly saw the validity of Mueller's point of view and was determined to reorient MSFC along systems engineering lines so as to better coordinate MSFC's design efforts.
>
> Tompkins investigated MSFC's organization and soon concluded that the design laboratories were overly oriented toward "low-level subsystems engineering." As one manager put it, "If we had a lawnmower capability at the Marshall Center, we'd put lawnmowers on all the vehicles." To combat this, Tompkins recommended significant strengthening of the systems engineering office. With this change, systems engineering by late 1968 became

a much stronger element within MSFC, albeit weaker in the traditional rocket groups than the newer organizations that focused on other projects. As MSFC's "family" organization and expertise in rocketry grew less important, systems engineering took their place. Formal coordination processes replaced the informal methods that sufficed in von Braun's heyday. (Johnson 2002, 150–151)

Johnson, who believes that systems management, including systems engineering, was the "secret" of the success of the Apollo Program, asks why NASA's most experienced engineers took so long to embrace systems engineering. His answer is the long continuity of the team: "Simply put, when each team member knew the job through decades of experience and knew every other team member over that period, formal methods to communicate or coordinate were redundant" (151–152). In other words, Johnson is saying that effective communication explains why MSFC could work successfully without embracing systems engineering. That is an explanation we must reconsider in analyzing the *Challenger* and *Columbia* accidents in later chapters.

On September 16, 1968, I briefed von Braun and Shepherd about the reorganization and the future role of systems engineering as described above. The meeting took three hours. We were interrupted when Bonnie Holmes, von Braun's secretary and "boss" walked in to report that James E. Webb, NASA Administrator, had just publicly announced his resignation and indicated von Braun was free to comment. Von Braun dictated a statement, which she promptly typed up and returned to him. He asked her to give it to me. He saw that I was puzzled and asked me to check the grammar and style. I did so and pronounced it to be good English, but there was a statement that alluded to some "storms" through which Mr. Webb had steered NASA safely. "What does that mean?"

"The 204 fire, Governor Wallace, and the Alabama environment," he replied.

Getting back to the briefing, I sketched the approaches to reorganization that I had considered. After I had indicated which scheme I favored, von Braun pulled a penciled organization chart out of his jacket pocket and said, "We're close." His schematic was quite similar to mine. He said this was due in part to the influence of my recommendations in 1967. For example, I had made a recommendation in 1967 about reorganizing the staff and administrative offices. This change was now on both of our organizational schemes. Systems engineering appeared in a more authoritative position on both charts. Both schematics included a prominent new line office labeled "Science," which was an attempt to ameliorate the science-engineering communication barrier I had discovered the year before.

I also gave him recommendations on several other problems he had asked me to investigate. (Again, the curious reader can find a discussion of these matters in *Organizational Communication Imperatives.*) The long meeting ended with von Braun requesting that I not reveal his reaction to my proposed reorganization scheme. He wanted to make the final decisions, obtain Washington's approval in principle, and announce the changes himself. This made sense. I had hoped that the rest of week could be spent at a more relaxed pace in preparation for my drive north to a new job at Kent State University in Ohio. But many of the people I'd interviewed that summer wanted to know how my briefing with von Braun had gone and refused to take no for an answer, even when I explained my promise not to reveal von Braun's responses to my proposals. Therefore, the week was as hectic as any other as I ran around the center, briefing important people on my findings and recommendations but keeping von Braun's reactions and intentions confidential.

Cultural Categories

Let's take a break from my description of the Marshall Center in the Sixties to see how it matches up with McCurdy's cultural categories as amended in Chapter Three.

Research and Testing

Not much needs to be added. This was a deep cultural characteristic of the Marshall Center even after the build-up of project management and the transition from a do-it organization to a see-it-is-done organization.

In-House Technical Capability

The arsenal concept of in-house research, development, fabrication, and testing made in-house technical depth a central dimension of the Marshall Center culture during the first decade of its life. My research in 1967, however, revealed that changes were underway that would weaken this aspect at the level of the laboratory. Informal conversations with new friends and acquaintances at the Marshall Center had brought to my attention that there was an apparent discrepancy between lab directors and workers in the perceived need for support contractors. The workers perceived the contractors' employees working beside them (at much higher salaries) as unwanted and overpaid interlopers who were doing work that should have been done "in-house." The lab directors felt that the contractors' employees were necessary to some extent, but it seemed to me that their presence was perceived as a status symbol by the lab directors; the larger the number, the bigger the symbol.

I briefed von Braun and his associates about this problem. I had been anticipated by H. L. Nieburg's observations that NASA headquarters forced von Braun into "arranging with contractors to recruit technical people who were placed on inflated, government-reimbursable company payrolls while assigned to work at Huntsville side by side with civil servants" (1966, 232). Despite the cultural bias for in-house depth in Huntsville, many in Washington believed that money given to contractors from around the country would broaden the Congressional support for NASA and the space program. It was clear to me that if the personnel at MSFC had had their way, they would have enjoyed an even greater depth of technical personnel. They reluctantly accepted the reductions in force, known to the government as RIFs, that cut into their technical strength.

Hands-On Experience

I found at MSFC in 1967 and again in 1968 a highly valued belief that an engineer with clean hands got no respect, as Rodney Dangerfield used to say. The norm was they had to get their hands dirty, touching the hardware, testing it, and fixing it. As mentioned in the previous chapter, engineers who became paper-shuffling managers often lost status even though they gained additional income. I was surprised to discover that part of space technology and space flight was dependent on the applications of art as well as science. Skilled technicians were considered artists necessary to the program. Welders, for example, get their hands dirty; welding is an important skill in modern society, but stop for a moment to consider the greater difficulty when the welder has to work with highly exotic metals of rocket engines and with space capsules that will be subject to unearthly high temperatures. Welders and other technicians have to get their hands dirty; so it is with engineers as well. McCurdy's observation about the early culture of NASA is supported by observations about this dimension at MSFC.

Exceptional People

Like McCurdy, I found this cultural characteristic of the OTC to be salient at MSFC in the late Sixties. The only exception was my realization that it was difficult for them to recruit African Americans with a technical background to live in Huntsville, Alabama. Otherwise, the personnel were outstanding. Perhaps the best way of demonstrating that McCurdy's cultural dimension applied to MSFC is to report that because of its size and the quality of its engineers, the Huntsville group was considered by other NASA centers to be intellectually arrogant, even difficult to deal with in that regard, and not so much given to open communication with perceived inferiors as they were with colleagues at their home center.

Risk and Failure

On January 27, 1967, just six months before I arrived for work at the Marshall Center, a full-scale test was prepared for the Saturn V, the moon rocket, with three astronauts sitting on top of it in the space capsule on the launch pad at Cape Canaveral (Apollo 204). There was no intention to launch the rocket; the purpose was to fill the capsule with oxygen and simulate the launch. A fire somehow broke out, killing the three astronauts, Virgil "Gus" Grissom, Roger Chaffee, and Edward White, before technicians could get them out. Although the responsibility for the capsule was with the Johnson Space Center in Houston, personnel at the Marshall Center were still shaken when I arrived six months later. It was the gloomiest period of the Apollo Program. Investigations were underway, and although the Marshall Center received praise in the report, it was a grim reminder of the risks involved in space travel. Informal communication among the Marshall engineers that I picked up contained criticism of the lack of in-house technical depth at JSC and of the contractors they supervised.

Frontier Mentality

The first love of the German team was the exploration of space. They developed the rockets used in the programs summarized by

McCurdy, moving from challenge to challenge, living the dreams they had had in Germany. I was told by his German colleagues that von Braun had sketched the moon rocket on the back of an envelope at Peenemünde. He and his colleagues had the vision of a space station and a winged rocket to service it, as well as a vision of manned flight to Mars.

Open Communication

In addition to the channels and practices I've already described in the communication culture of the Marshall Center, perhaps another manifestation of openness is the fact that a 33-year-old Summer Faculty Consultant in Organizational Communication was brought on board and encouraged to get his "hands dirty" by searching for problems. I was turned loose to find out whether the earthquake prediction system was working.

I was asked to return to the Marshall Center in 1969 but had received a travel grant from Kent State University to spend part of the summer in Ireland doing research on James Joyce. Again, in 1970 I was asked to return for the summer to work on organizational problems having to do with systems engineering. My plans for that summer were changed on May 4, 1970, when four students at Kent State were killed and nine wounded by the Ohio National Guard during a protest over President Nixon's incursion into Cambodia during the Vietnam War. I was asked by the President of the University to serve on a Commission investigating the shootings. This work led to a book, *Communication Crisis at Kent State: A Case Study* (1971). (My co-author was an instructor at the university named Elaine Anderson; since 1971 she has been Elaine Tompkins, and thereafter we told students to be careful in selecting co-authors!)

After I got back from Ireland in 1969 I gathered with friends to watch the moon landing on the evening of July 21. We were, like most Americans, keen with anticipation, but few people watching were as identified with the project as I. The Saturn V had performed flawlessly in placing the astronauts on the precise trajectory toward the moon. Now the tiny spacecraft had to land them on the lunar surface and bring them back to a rendezvous with the larger spacecraft orbiting the moon. Then we got those first eerie sounds and pictures from the moon.

Armstrong: "Houston, Tranquility Base here. The Eagle has landed." Those words broke the tension for "some sum of billions of eyes and ears around a world which had just come into contact with another world" (Mailer 1971, 335). None of us at the party could understand Armstrong's words when he first stepped on the moon, a serious rhetorical disappointment, but we read them later. Norman Mailer (1971) got them down and added a terse comment:

> "That's one small step for a man," said Armstrong, "one giant step for mankind." He had joined the ranks of the forever quoted. Patrick Henry, Henry Stanley and Admiral Dewey moved over for him. (352)

We had redeemed Kennedy's promise, and the nation's mood seemed to be one of regained pride. Some critics argued, however, that it was nothing more than a publicity stunt to distract us from protesting the war in Vietnam. Others complained that it was a waste of money that could have been better spent on social programs. (I worried about that argument until I became convinced over the years that cuts in NASA's budgets didn't correlate with increases in social programs.) The most incredible reaction to the moon landing came from a large group of American citizens, numbering perhaps in the millions, who *didn't believe* the landing had been accomplished. They believed instead that all the pictures came from a televised simulation on some sound stage. I remember vividly the woman in Florida who expressed this belief with impeccable logic. She said she couldn't pick up television stations in New York, so how could she possibly be getting pictures from the moon?

I would later begin an article about MSFC and the Apollo Project with these lines:

> The Apollo Project was the greatest technological achievement of our society. It worked. During an era of demonstrations, violence, war, assassinations, riots, scandals and technological failures—it worked. It was also the largest engineering project in history (Tompkins 1977, 1)

I also described the ways in which Apollo was a success in terms of organization-as-communication. But that night in 1969, I went to bed happy with the knowledge that the United States could do something right, even noble, and that the Marshall Center had played a major role in the moon landing. Although my role had been tiny, I felt as if I had been part of the magnificent achievement.

Topoi *and Tradeoffs*

One more concept must be introduced under the heading of *topoi* and tradeoffs. *Topoi* is the Greek plural for *topos* (the singular), an idea of Aristotle's, the idea of a place where one might look for premises for lines of argument. I discovered that there were three *topoi* at the Marshall Center: *reliability, time, and cost*—and in that order of importance. That is, if two engineers disagreed about a problem or a solution, one could appeal on the grounds of cost—my solution is cheaper. The other could say mine will take less time and we can keep on schedule. A third could then play the trump card by saying my solution is more reliable, safer for manned space flight. Reliability or safety was the *primus inter pares,* or first among equals, among criteria or places to look for arguments. The difficult decisions were those in which all three *topoi* had some traction, some ground under their feet. In those cases the only resolution could be a tradeoff, a kind of compromise among the three lines of argument. As an internal footnote, von Braun told me on several occasions when he sent me out to help settle disputes, "What I most want to avoid is a foul compromise." Not a compromise, but a *foul* compromise. As I understood the difference, a foul compromise would be one in which the first *topos,* safety, was compromised, or one in which a decision might be made to satisfy someone's ego. It was patently clear that *reliability* or *safety* had precedence over the other two criteria.

Let me also say something about language as an important part of the communication culture of NASA. As mentioned above, von Braun warned me about the dangers of alphabet soup. His assistant, Jim Shepherd, gave me special problems to research, one of which was the briefing room in which project managers conducted their briefings. The acronyms and displays were so baffling that he couldn't understand them. I got behind the scenes and exposed some of the practices as inaccurate and misleading. As noted above, Von Braun stopped attending the briefings in the Control Center as a result. I see now that Kenneth Burke's classic essay, "Terministic Screens," is a powerful tool in understanding culture. Burke's argument is that the separate academic disciplines have different nomenclatures, each of which calls attention to certain phenomena and screens out others. At the Marshall Center we had to deal with all of the nomenclatures of the engineering disciplines and physical sciences. One of the serious problems in communication I discovered there was what I called the science-technology barrier.

Engineers who keeps their hands dirty are functioning in a world of reliability, time, and cost. They are concerned with hardware and successful missions. They write "state-of-the-art" papers for other engineers. They talk in acronyms called alphabet soup. Physical scientists are interested in publishing papers in journals; their time frame is a lifetime, a career in pursuit of a Nobel Prize. They sometimes look down on the practical work of the engineer.

How important in retrospect the Monday Notes seem as an integrating communication device. The requirement of hearing from all the disciplines, including the management jargon, forced all parties to learn the different technical dialects and their different assumptions and premises. No single dialect was allowed to dominate and screen out others. As Weick and Ashford said of the Monday Notes, they regularly made people see that there were different ways to frame common problems. This helped a strong culture to become a learning organization at the same time.

As explained above, during the summer of 1968 von Braun asked me to tackle the problem of reorganization, and after pursuing the assignment I was persuaded by the need to place Systems Engineering in a more central role in the organization for the reason that its job is to speak in a language that integrates the specialized subdivisions. Again in retrospect that recommendation, plus the neutralization of the Control Center, prevented or at least delayed the ascension of a management nomenclature that could screen out other competing dialects.

More could be said about open communication, communication as organizational learning, communication as culture at MSFC during the first decade of the center's life, but perhaps this is enough to justify making this a separate and overarching dimension of the culture.

Organizational Identification

The construct of organizational identification is one with which I am identified; it has its origins in my experience at the Marshall Center. I noticed the pride that NASA employees took in their work, their lab, their space center and its missions and projects. They had models of NASA rockets on their desks, photographs of rockets on their walls. They worked long hours without extra pay. They looked for new tasks and new responsibilities not assigned to them. They exhibited many manifestations of their commitment; one could see this dimension of culture in their actions; it was palpable. I could both see and infer their identification with the culture by working with them. Identification is an attitude or incipient action to Burke, explaining why he placed the theory under his general heading of motives.

I had met Burke at Purdue University five years earlier, and in preparation for Burke's lecture, "On the Suasive Nature of the Most Ordinary Nomenclatures," now known in print as the essay referred to above as "Terministic Screens," I learned that Burke's larger program dealt with motives and motivations with three subheadings: a grammar, rhetoric, and symbology of motives. Burke's *A Rhetoric of Motives* (1950) was devoted to the process of rhetoric, or persuasion. The traditional approach to persuasion was the paradigm case of a speaker seeking to change the attitudes and actions of an audience. One could later add writers and readers to speakers and listeners.

Burke realized that in the twentieth century there was need for a new paradigm of persuasion, the case in which *no one is trying to persuade us.* In a retrospective look at his own book on rhetoric 22 years later, Burke summarized his ideas in this way:

> [We] spontaneously identify ourselves with some groups or other, some trends or other—and we need a term for this kind of persuasion in which (you might say) we spontaneously, intuitively, and often unconsciously, act upon ourselves. (Burke 1972, 27–28)

This acting-upon-ourselves is processual, the manner in which we come to adopt the values and attitudes of a larger group. Once we have accepted these values and attitudes we use them as premises in making decisions. The girl who wants to become an astronaut has persuaded herself to become part of something larger, having done so by the process of organizational identification; she has acted upon herself to create her new identity and to act upon that identification.

I could see the validity of the theory by observing others, but in addition could feel changes in me. By doing communication research in the Marshall Center, I found myself using the first-person plural pronoun *we* in reference to the space center and NASA, its people, practices, and values. I put a photograph of a rocket on the wall of my university office, carried a clipboard with the NASA logo on it, and brought up the organization and its projects in conversations and lectures. Finally, I made a more thorough study of Burke's works and organized in 1971 the first empirical study of organizational identification (Tompkins et al. 1975).

Unobtrusive Control

This section is not intended as advertisements for me so much as an introduction to ideas that will be applied in a later chapter trying to understand the *Columbia* tragedy. At Purdue University in the 1980s, I worked with outstanding graduate students, among them Connie Bullis (e.g., Bullis and Tompkins 1989; Bullis 1991) and George Cheney (e.g., Tompkins and Cheney 1983, 1985; Cheney and Tompkins 1987), who chose to work with me to advance a research program in organizational identification. The insights of Herbert Simon, Nobel Laureate in Economics, were combined with those of Burke; case studies and quantitative research improved our understanding of organizational identification. But I had noticed during my work at NASA's Marshall Center that identification worked as a form of control as well as of motivation and identify formation. That is, if people identify with an organization, its goals and premises, managers need not worry so much about motivating and directing them. People control themselves in such circumstances.

In 1985 the job of integrating the theory of organizational identification into a new theory of control was accomplished: "Communication and Unobtrusive Control in Contemporary Organizations" (Tompkins and Cheney 1985); this work introduced the concept of "concertive control." Concertive control is not so much control by a hierarchy or supervisor as it is group self-control; workers decide how to do the work, then monitor and correct each other when necessary. A synonym is teamwork, in which there is intensive face-to-face communication and democratic decision making based on the larger organization's decision premises. I had noticed at MSFC that teams worked intensively within laboratories and in "working groups," teams with members from many different organizational units; the members identified with their teams as well as with larger organizational units.

After we moved to the University of Colorado in 1986, George Cheney—now a faculty colleague—and I worked with some excellent graduate students to press on with the theory and research into identification and concertive control. James Barker, for example, studied a manufacturing company in Aurora, Colorado, and published an article in the *Administrative Science Quarterly*: "Tightening the iron cage: Concertive control in self-managing teams" (1993); the article received from the Academy of Management an award as the Outstanding Publication in Organizational Behavior for 1993. A study by Barker and Tompkins (1994) found that some workers identified more with their work teams than with the larger organization. Loril Gossett (2003) pursued identification and control among "temp" workers. Gregory Larson pursued the concept of unobtrusive control in an aerospace contractor that had done work for NASA. This research provided new insights into NASA's changing culture over time. A paper by Larson and Tompkins (2003) dealing with the same aerospace company and its connection with NASA completes the circle, returning to the organization in which I was inspired to begin my work in communication, identification, and control.

But there is a dark side to identification and control. One can only assume that Nazi underlings identified with Hitler and his programs. In addition, managers can introduce concertive control and encourage identification in such a way that the people involved do not understand what is happening to them. We, Tompkins and Cheney, wrote the first draft of the theory, at least in part, to alert people to their situation. Burke, citing Aristotle's *Nichomachean Ethics*, believed that truly human action is *fully informed*. For that reason, the theory of identification and control has had from the beginning a critical edge, so much so that the textbook we have used as a benchmark so far, the one that regards identification as one of the three most important constructs, separates it from the theory of concertive control. That text presents it with two other critical approaches and under the heading of "Power, Hegemony, and Concertive Control." Their reason is consistent with the original conception of the theory: Concertive control increases the total amount of control in the system because each worker is a supervisor—everybody is a boss—and the approach "may serve to oppress workers" (Modaff and DeWine 2002, 100).

MSFC during the Apollo era was a large and extremely complex organization working with dangerous materials and a high risk of disaster, yet they managed to put Americans on the moon and bring them back home safely. They did this during an era of protests and demonstrations against the Vietnamese war and the assassination of President Kennedy, his brother Robert, and Martin Luther King, Jr. It was the biggest engineering project in human history. It worked. They did it with brains, a frontier mentality, a high degree of organizational identification, and open communication. The culture was the closest approximation of the Ideal Managerial Climate I have ever seen.

This completes our description and analysis of culture and communication during the golden era of NASA. The next chapter will move forward to 1986, the year in which NASA and the nation were shocked and saddened by the tragedy of the shuttle *Challenger*. ✦

Chapter Five

The *Challenger* Accident

On Tuesday, January 28, 1986, 17 years before *Columbia* disintegrated, *Challenger*, a sister ship, exploded during liftoff at 11:38 a.m. Millions were watching on television, including schoolchildren across the country, because this was the flight to put a teacher in space. Seven astronauts were lost that day (Greg Jarvis, Ron McNair, Ellison Onizuka, Judy Resnik, Dick Scobee, Mike Smith, and the teacher-astronaut, Christa McAuliffe). NASA had been a relatively untarnished institution, highly esteemed in the public mind until then. The only other fatal accident was the fire that claimed three astronauts in 1967. The mystery of what happened to *Columbia* can't be solved without an understanding of what happened to *Challenger*.

George Cheney studied the mass media accounts and interpretations of the *Challenger* tragedy. In the days and weeks following the event, the accident was the lead story on television, a front page and cover story in newspapers and magazines. The videotape of the explosion was seared into the minds of Americans as they watched it being replayed over and over again. The word *tragedy*, Cheney found, was used over and over in commentary about the accident, and so he turned to Kenneth Burke's brilliant treatise on communication, *Permanence and Change,* for a better understanding of tragedy. Burke wrote of the "close connection between tragedy and purpose. We might almost lay it down as a rule of thumb; where someone is *straining* to do something, look for evidence of the tragic mechanism" (Burke 1935/1984, 195).

111

Cheney also found a linkage between *sacrifice* and *purpose* in the discourse about the accident. It was dramatized "in at least four ways:

> (1) the visual portrayal of the space shuttle's explosion, (2) the expression of personal identification with the *Challenger* because of the teacher-astronaut Christa McAuliffe, (3) the appeals to America's collective "destiny" in space, and (4) the conveyance of religious significance surrounding the disaster. (Cheney 1987, 3)

I followed the *Challenger* accident with numbness and a yearning for what had been lost. I certainly felt a personal sense of loss over the tragedy because of my past experiences with NASA and the Marshall Space Flight Center. I had been part of the organization; it was part of me. At the time of the *Challenger* tragedy I was a professor in the Department of Communication and associate dean of the School of Liberal Arts at Purdue University. Purdue proudly claimed to have graduated more astronauts than any other university. Neil Armstrong, the first man to step on the moon, was a Purdue graduate; so was Gus Grissom. We identified with NASA and the space program, and we agonized over the accident.

The *Report of the Presidential Commission on the Space Shuttle Challenger Accident* became available on June 6, 1986. President Reagan had appointed the group, and it became known as the Rogers Commission because it was chaired by former Secretary of State, William P. Rogers. On page five of the preface to the report there is the statement that after the landing of the *Columbia* shuttle on July 4, 1982, NASA "declared the space shuttle 'operational,' a term that has encountered some criticism because it erroneously suggests that the shuttle had attained an airline-like degree of routine operation." *(Could the word* operational *be another example of the Whorf hypothesis, that is, a word that affects behavior? Even the Commission believed that NASA had been "straining" after its purpose, the mechanism underlying tragedy.)*

The Rogers Commission Report methodically rejects all hypotheses save one: there was no evidence of sabotage; the external tank hadn't contributed to the cause of the accident; the main shuttle engines didn't contribute to the cause; nothing in the Orbiter/payload interface contributed to the accident. The cause was a failure of the pressure seal [O-ring] in the right solid rocket motor. The technical cause of the explosion had been identified. (An O-ring is like a huge

washer used in the joints of the solid rocket booster to prevent searing gases from escaping.) Chapter V of the report, "The Contributing Cause of the Accident," sought to identify the human failures
behind the technical failure. The opening paragraph of the chapter
begins:

> The decision to launch the *Challenger* was flawed. Those who
> made that decision were unaware of the recent history of prob-
> lems concerning the O-rings and the joint and were unaware of the
> initial written recommendation of the contractor advising against
> the launch at temperatures below 53 degrees Fahrenheit and the
> continuing opposition of the engineers at Thiokol after their man-
> agement reversed its position. They did not have a clear under-
> standing of Rockwell's concern that it was not safe to launch be-
> cause of ice on the pad. If the decision makers had known all of the
> facts, it is highly unlikely that they would have decided to launch
> 51-L on January 28, 1986. (82)

That paragraph needs some unpacking for people who didn't
follow the case carefully, or for those whose memories of the events
have faded in the past 17 years. First, someone failed to *communicate*
the facts prior to launch. Those who made the decision didn't know
that Marshall Center engineers had recently been working on the O-
ring problem. Nor did they know that the initial written recommendation against the launch was made by the contractor, Morton-
Thiokol, during a teleconference on Monday, January 27, 1986, the
night before the launch. Representatives of three organizations
were on the telephone line from the Marshall Center, the Kennedy
Center, and Thiokol, the shortened name of the private contractor
that made the solid boosters. The Thiokol engineers recommended
against the launch the next day because icicles had formed on the
launch stand in Florida, and they had no experience with the O-
rings below 53 degrees F.

Testimony about this telephone conference (with support from
graphics and written messages via fax) represented Marshall Center personnel as reversing their usual approach in dealing with the
personnel of Morton-Thiokol. Instead of pressing the contractor to
prove that *Challenger* would fly, they insisted that Thiokol back up
its recommendation by proving it would *not* fly. This was a complete reversal of past practices, inconsistent with the philosophy
and practices of communication, penetration, and automatic responsibility, including the *presumption* that no "manned" mission

should take place if there were any reasonable doubts about the safety of the mission. The burden of proof was always with the contractor to prove the flight would be safe.

The *Challenger* teleconference reversed the established "burden of proof," a concept originally defined by the English rhetorical theorist Richard Whateley (1787–1863). Burden of proof is a dialectical term whose meaning is made clear by its opposite, the presumption. Presumption refers to the beliefs, attitudes, and institutions currently accepted—the status quo. The status quo will continue without defense until sufficient reasons are adduced against it to produce change. The side or forces that seek such changes must generate sufficient reasons for change. This is a natural form of presumption evident in all aspects of life. The presumption—the basic value premise—in NASA had always been that manned space flight should not be undertaken when there were any reasonable doubts about the safety of the mission. The presumption of NASA was that space flight was inherently dangerous. To counter this presumption, the burden of proof demanded that sufficient reason and evidence be cited to prove beyond all reasonable doubt that the spacecraft will fly safely.

The reversal of this approach in the case of *Challenger* was certainly not an original observation on my part; many researchers in the field of organizational communication came to the same conclusion. Nor was it lost on the Rogers Commission. My concern, however, was somewhat different; I knew that this event also reversed the past practices of the Marshall Center. Wernher von Braun, the director of MSFC from its inception until 1970, was once asked by a congressional committee why contractors performed better for NASA than for other clients. Von Braun's answer was, in a word, *penetration.* What he meant was that MSFC engineers got inside the contractor organization to look for potential problems; they forced the contractor personnel to prove the rocket was going to be safe.

As we shall see, another commentator took a different stance on this matter, but it should be repeated that the Rogers Commission made this flawed process of communication a central contributing cause to the ill-advised launch. I did the same in my 1993 book about the accident. Now we can add to that list the *Columbia* Accident Investigation Board (CAIB) Report released in late August of 2003. Looking back on the *Challenger* accident and the Report of the Rogers Commission, CAIB asserted the following:

The Rogers Commission concluded "the decision to launch the *Challenger* was flawed." Communication failures, incomplete and misleading information, and poor management judgments all figured in a decision-making process that permitted, in the words of the Commission, "internal flight safety problems to bypass key shuttle managers." As a result, if those making the launch decision "had known all the facts, it is highly unlikely that they would have decided to launch." Far from meticulously guarding against potential problems, the Commission found that NASA had required "a contractor to prove that it was not safe to launch, rather than proving it was safe." (CAIB Report 2003, 100)

To understand the impact of this reversal of the burden of proof, consider how the concept is used in our judicial system. A person in the United States is presumed to be innocent (the presumption) until proven guilty (the burden of proof). The defendant is presumed innocent but that isn't the same as assuming or believing the person is innocent. The defense has the presumption; the prosecution has the burden of proof. The prosecution must muster enough proof to establish beyond a reasonable doubt that the defendant is guilty of committing the crime.

Why does our system *artificially* establish that the defendant has presumption and place the burden of proof on the prosecution? Our system correctly recognizes the possibility, even inevitability, of human error in the judicial process and in the collection of relevant evidence. Based on our belief of the inherent value of an individual life, the court system accepts that it is better to let a guilty person go free than to incarcerate or execute an innocent person. Think about this in a personal sense, one close to home. It might be impossible to prove your innocence of a crime such as murder or of misconduct such as cheating on a girlfriend, boyfriend, or spouse. If you were at home alone, sleeping, with the phone off the hook, it would be extremely difficult to prove where you were—or weren't. So the burden of proof is demanded from the prosecution in order to convict a person of a crime.

The safety issue is analogous. The assumption in NASA had been that in manned flight, safety was the preeminent value premise, or *topos*. Those at NASA knew that blasting persons into space on top of a man-made rocket was inherently dangerous, a high risk. For that reason the rocket was presumed to be unsafe. Before a launch the decision to "go" had to be based on arguments sufficient to establish beyond a reasonable doubt that the rocket was safe. They did not presume it was safe; it takes more argument and evidence to establish the burden of proof than to rely on presumption. When it came to safety, they required the rigorous demands of the burden of proof before making a decision to launch, thereby increasing the probability of a safe mission.

As I read the Rogers Report, I noticed that the description of many communication events didn't fit with the philosophy and practices I had observed and written about, including the Monday Notes, penetration, automatic responsibility, and the emphasis on the upward-communication of problems. What had happened to the Marshall Center? I began to develop a hypothesis of organizational forgetting, amnesia. In order to test—in the old Marshall tradition—the hypothesis of institutional forgetting, I returned to the Center in order to interview 16 of the top managers on January 9 and 10, 1990, four years after the *Challenger* accident.

To bring me up to date for the interviews, Marshall sent biographies of their current managers and the *Strategic Plan Progress Report: A Reflection on 1988.* The latter taught me that the number of employees was 3,240, fewer than half the number during the Apollo Program. The report said the Marshall Center was committed to excellence and to *"penetrate* our projects in more technical depth" (emphasis added). The last page of the report listed the Marshall Center's "Guiding Principles," the last of which was "Communications—[to] assure open and effective communication with our people, government and industrial partners, and with the public." Open communication. *(Perhaps they haven't forgotten, I thought.)*

The managers opened up to me immediately. They were highly critical of the two previous Directors, Rocco Petrone and William Lucas. Lucas resigned after the *Challenger* disaster, as did the Directors of the Kennedy Space Center and the Johnson Space Center. The interviewees were in general pleased with their current administrators and the communication culture. There were, however, some clear indicators that time had changed the culture. The reduced workforce made penetration of the contractors more difficult, what with fewer people to monitor more projects; the diversification of projects since Apollo made it harder to practice automatic responsibility because it was more challenging to have technical competence in so many different areas. Most disappointing, only four of the 16 managers could define automatic responsibility correctly.

The Monday Notes had been transformed into the Weekly Notes and there were other changes as well. Von Braun's practice had been to ask for hard copy so that he could write in the margins of each note before reproducing and returning the entire package of notes to each contributor. In 1990, the director no longer wrote in the margins; instead, notes were communicated up the line via electronic mail. The loss of open feedback from the boss may also have been a matter of the new director's style, but we can cite this as one of the inevitable tradeoffs involved in adopting new communication technologies.

In addition, the bypassing feature had been abandoned by allowing the intervening level of management to screen the messages. The methods of generating the notes were now varied, and although the interviewees were positive about the Weekly Notes, it was clear to me that the system no longer had the same power and discipline, was no longer *"a rigorous and regularly recurring discipline of communication"* within the organization. Apart from the Weekly Notes, it seemed clear that there was a fairly serious degree of organizational forgetting over the years, as well as an erosion of strength in some of MSFC's more traditional communicative practices. I hadn't planned to ask about communication as a contributor to the *Challenger* accident, but because of answers the managers volunteered to other questions, it was natural to follow up by asking six of the managers: Did ineffective communication contribute to the accident? Two typical answers:

> The teleconference got turned around. When challenged, Thiokol, rather than standing behind their data, told the government what they wanted to hear. . . . Yes, there was a communication problem.
>
> My impression is that Thiokol was opposed to the launch at the engineering level. They were surprised at the stance of the Marshall managers. Always before, the Marshall managers would make sure that it was okay to launch. The teleconference was atypical. We needed more openness in the agency than we had then. We've seen the results otherwise.
>
> I asked, "What results?"
>
> The manager said, *"Challenger."* (Tompkins 1993, 167)

In retrospect one can say that the reversal of the burden of proof was "atypical," as the manager put it—a flat-out deviation from the Marshall culture that surprised the Thiokol managers.

In addition, another manager answered the question in this way: "Was it a factor? My opinion is that it is probably a true statement. We developed a feeling, 'Well we've had 25 flights and we weren't going to have a failure.' We ignored or put off the problem." *(In retrospect this may represent a case of hindsight bias on the part of the mission managers.)* Because the shuttle had flown safely 25 times, their hindsight bias wouldn't allow them to analyze clearly the signs of O-ring erosion, wouldn't allow them to see that something not expected by the design was an antecedent to disaster.

There were other books about the *Challenger* disaster. McCurdy's *Inside NASA* appeared in 1993. On the tenth anniversary of the disaster two other important books appeared: in alphabetical order, Claus Jensen's *No Downlink: A Dramatic Narrative about the Challenger Accident and Our Time*; and Diane Vaughan's *The Challenger Launch Decision: Risky Technology, Culture, and Deviance at NASA*. Vaughan's volume received more attention than the others and is considered an important reference work on the subject. Both books were reviewed by me with help from Kurt Heppard (an Air Force officer and faculty member in the Department of Management at the Air Force Academy) and Craig Melville (doctoral student at the University of Colorado); this discussion is a shortened version of our review article that appeared in *Organization*, an interdisciplinary journal of organization, theory, and society, in November of 1998. The review article makes a close comparison between the Jensen and Vaughan books, using McCurdy and other sources when needed.

The books by Jensen and Vaughan are quite different. Claus Jensen is described on the dust jacket as a professor of literature who lives in Aarhus, Denmark. Diane Vaughan is a sociologist at Boston College. There is a tension between the two works, a difference of emphasis and argument. I shall try to show the contrasting theses they advance and will search for common ground or a way of resolving their differences.

The two books develop different explanations for the ill-advised decision to launch the shuttle in unprecedented cold weather on January 28, 1986. Jensen highlights the power and potential of the individual to protect the system from itself when deviations from the norm occur. In his view, individuals can be heroic—or at least responsible—when an unexpected problem threatens the system. Vaughan's attention, by contrast, is fixed on the evolution and power of social structures; in the case of *Challenger*, she argues that structures evolved that led to the "normalization of deviance" in NASA and against which individuals were either irrelevant or impotent.

Claus Jensen and *No Downlink*

Claus Jensen grew up in Denmark and, as he explains in his preface to the book, Americans were known to him as a boy primarily as the liberators of Europe. When he and his parents watched the tiny dot that was *Sputnik 1* moving against the dark autumn sky, he knew the Russians were ahead in the space race. But Jensen had an uncritical faith in the Americans, developed by studying well-thumbed copies of *Popular Mechanics* in his local library. He had no doubt his "men from *Popular Mechanics*" would catch up with and outrun the Russians. His initial understanding of the accident was that it was an inevitable technical foul-up. He changed his mind when he read Richard Feynman's account, particularly when Feynman visited the Marshall Center as a member of the Rogers Commission and asked to speak to three engineers who worked on the boosters. Let's retell the story of Feynman's two experiments.

Feynman's Two Experiments

The first experiment. Despite Richard Feynman's Nobel Prize in Physics (1965) and his best-selling book, '*Surely You're Joking, Mr. Feynman,*' he is best known to space buffs for a simple but dramatic experiment he conducted with a rubber washer and a cup of ice water. The Rogers Commission heard testimony about the O-ring, a rubber gasket that was supposed to contract and expand within a "field joint" (so called because the boosters were assembled in the field) to prevent dangerous leakage of hot gases from the solid rocket boosters. During the infamous teleconference, the decision about whether or not to recommend a launch hinged partly on the weather at the Cape and its potential effect on the O-rings.

On the morning of Flight 51-L, the weather was colder than at any previous launch of a shuttle, and some engineers argued that they were, therefore, outside the "envelope" of experience. Icicles were hanging from the launch tower, and could have been a second threat to the flight if sucked into the rocket engines. Would the cold weather affect the performance and the resiliency of the O-rings to expand and contract? Some thought so, others thought not.

The Rogers Commission, as mentioned, took testimony on this issue and made its conclusion crystal clear. Before doing so there was a climactic point in its hearings in which a model of the field joint, complete with the inserted O-ring, was passed from panelist to panelist. Feynman had made a trip to a hardware store to prepare for this moment. When the model reached him, he pulled out a clamp and a pair of pliers. Feynman then extracted the O-ring with the pliers, squeezed it with a clamp, and plunged it into a cup of ice water. He announced the results of his experiment:

> I take the clamp out, hold it up in the air, and loosen it as I talk: "I discovered that when you undo the clamp, the rubber doesn't spring back. In other words, for more than a few seconds, there is no resilience in this material when it is at a temperature of 32 degrees. I believe that has some significance for our problem. (Feynman 1989, 151–153)

This strikingly simple, dramatic experiment—the OTC tradition of testing—demonstrated that cold weather was bound to affect the performance of the O-rings. The Morton-Thiokol engineers who had raised doubts about the O-rings had been right.

The second experiment. It was Feynman's second, less-well-known experiment that changed Claus Jensen's mind. When the time arrived for Feynman's meeting with the three engineers at the Marshall Center, the engineers' boss showed up even though he was uninvited; he justified his presence by saying he was an engineer with management responsibility—even though it was clear he was there to monitor the actions of the engineers. Impatiently interrupting a canned presentation, Feynman asked them all to write down on separate pieces of paper the probability of booster engine failure in a shuttle flight. The three engineers' probabilities were close: 1 in 200, 1 in 200, and 1 in 300. The manager refused to give an answer. After some intense prodding he came up with a number close to the estimate Feynman had received from another manager: 1 in 100,000. Feynman and the engineers were dumbfounded. How could there be such a difference? Feynman concluded that there was a serious communication barrier—or what I'd call *semantic-*

information distance—between the engineers and their managers, allowing the managers to be wildly optimistic about difficult technical issues. Jensen then began to think differently about the case and began to see the shuttle as a national symbol of the dark side of high technology and large, complex organizations.

The 'Challenger *Syndrome'*

We are surrounded in the modern world by ever bigger, ever more technically advanced organizations, says Jensen, and relatively small slip-ups or deviations from normal conditions can cause disastrous results—we might even expect them as in the "normal accident" described by Charles Perrow (1984). Jensen comes close to prophecy in this remark:

> Whereas, in the past, NASA's managers came from a technical background and had, therefore, great respect for their own technical experts and their judgment, these managers were now being supplanted ever more often by a burgeoning layer of bureaucrats, who had never wielded a monkey wrench and could only manage by means of orders, memos, rules and regulations. And it is not hard to imagine the kind of commitment and personal judgment that a technician is going to display in such an environment. *If that's the way the boss wants it, then that's what he'll get; it's his respon - sibility, not mine, thank goodness.* (363)

Jensen doesn't quite express it in the terms of cultural analysis, but he clearly implies that the bureaucratic managers who had never handled a monkey wrench, who had never got their hands dirty, represent a *culture different from the engineering culture.* He does grant that the system, the bureaucracy, can be powerful. An individual's own ethical instincts are often all he or she has to go by, and "great strength of conviction may be required for an individual to speak his or her mind" (370). When deviations threaten the system's social ethics and organizational success, individuals with personal integrity, intuition, and common sense must reject the logic of the organization in order to save it from itself. *The organization can't save itself.* At the conclusion of his narrative Jensen describes the "*Challenger* Syndrome," explaining that the innovative thinking, personal judgment, and responsibility so characteristic of the young NASA gave way over time to a bureaucratic mentality.

Diane Vaughan and *The Challenger Launch Decision*

By contrast, Diane Vaughan's account in *The Challenger Launch Decision* is eclectically steeped in sociological theory. She interviews the principals, employing what she calls "historical ethnography." The book and appendices are full of information about the fateful and fatal launch; for these reasons I recommend it as a mandatory resource for any serious student of the *Challenger* disaster.

Vaughan notes at the beginning that her background as a sociologist gave her certain preconceptions about the launch decision. She was inclined to believe that there had been managerial wrongdoing; after her exhaustive research, however, she came to a different understanding. She developed a theory of normalized deviance in which the shuttle program drifted into the inevitable disaster. She argues that this normalization of deviance was created by the culture of production and structural secrecy. Normalized deviance cycled through (1) some sign of danger, (2) an official act that acknowledged the escalated risk, (3) a review of the evidence, (4) an official act indicating the acceptance of escalated risk, and (5) a shuttle launch.

Managers, primarily middle managers, said Vaughan, followed procedural rules and acted ethically, and thus were incapable of overcoming a system that had spun out of their control—out of anyone's control. Vaughan concentrates on the "structure of power and the power of structure and culture" (xv). Her account challenges the widely held notion—which she refers to as the "conventional" explanation—that the 1986 *Challenger* disaster was largely the failure of the communication system in the NASA organization.

To better understand Vaughan's position, let's briefly review the conventional explanation: Within 120 days of the disaster, the Rogers Commission had found the technical flaws in the solid-rocket booster and the parallel organizational findings that there were serious problems of communication within NASA and in its relationship with contractor firms. Chapter Three of this book considered Howard McCurdy's analysis of the original technical culture presented in his 1993 book—research, testing, open communication, and so on. This was the culture that gave NASA its early and spectacular successes. McCurdy argues that the OTC gave way to a much weaker culture in the second and third decades of NASA's life

and that this weakened culture, including less-than-open communication, explains the failure of NASA in the *Challenger* tragedy.

My own book appeared in 1993, the same year McCurdy's came
out—and, as noted previously, we were unaware of each other as
we wrote. My description of culture and communication at the Marshall Center is consistent with McCurdy's during the first decade of
NASA's existence. I did expand the importance of "open communication" by equating it with the culture of the Marshall Center.
McCurdy and I were seeing and describing the same culture. After
the *Challenger* disaster I described a form of organizational forgetting—caused by a combination of new engineers, fewer people, and
new managers, some of whom were outsiders to the culture, and a
new diversity of projects. This analysis is similar to McCurdy's notion of a "weakened" culture. I also blamed deviant communication
practices during the fateful launch decision, including the shifting
of the burden of proof.

Vaughan's view rejects the conventional explanation given by
the Rogers Commission, Jensen, McCurdy, me, and other writers, in
order to emphasize the sociological thesis of the "normalization of
deviance," a phrase similar to McCurdy's OTC expression, the
"normalization of risk" and Perrow's phrase, the "normal accident." Vaughan also denies that the accident was caused by problems of communication, including the infamous shifting of the burden of proof in the teleconference on the eve of the fatal launch.

Vaughan's main reason for rejecting the conventional explanation dealing with communication is her analysis of the teleconference on the eve of the *Challenger* launch. George Hardy, Marshall's Deputy Director for Science, said he was "appalled" to hear
engineers at Morton-Thiokol recommend against launch on the
grounds that they had no experience with the suspect O-rings at
temperatures lower than 53 degrees F. Lawrence Mulloy, Marshall's
Manager, Solid Rocket Booster Project, then posed this question:
"My God, Thiokol, when do you want me to launch, next April?"
MSFC middle managers were aggressive in asking for more data
and perhaps demeaning in their strident demands for more quantitative data to support the no-launch recommendation from the
Thiokol engineers.

No contractor had ever made a no-launch recommendation—
even though that didn't change the burden of proof. The logical or
argumentative presumption had always been with safety; the bur-

den of proof always assumed by contractors was to prove that the rocket or shuttle could be launched safely. On that night before the launch Thiokol engineers and managers felt that the managers from Marshall required them to prove that they were *not ready to launch* (Jensen, 304; Vaughan, 41).

Vaughan's interpretation turns this episode upside down. She explains why Boisjoly and other Thiokol engineers *perceived* the burden of proof in this way:

> The Thiokol engineers believed that the burden of proof was different than in the past because *they had never come forward with a no-launch recommendation in an FRR [Flight Readiness Review] before.* Marshall managers and Thiokol proponents of the no-launch recommendation were both partially correct: managers did not behave differently, but the burden of proof did shift. It shifted because the contractor position shifted. . . . Since their position deviated from the norm, the burden of proof deviated from the norm. (Vaughan, 343, emphasis in the original)

This passage reveals an unusual understanding of presumption and the burden of proof. Instead of basing her version on traditional, logical, argumentative responsibilities, Vaughan thinks of the decision process in terms of social expectations. Contractors had always recommended a launch. If this contractor recommends against a launch, it will have to prove why. Vaughan's explanation may even make sense in a psychological sense. Nonetheless, this incident was by all accounts atypical of how NASA operated during its golden era. Safety no longer had the presumption. The tense battle between the managers and engineers also suggests people representing two different cultures.

A Faustian Bargain

The CAIB report on the *Columbia* accident shows that the old presumption was still true in 2003, because NASA has a "Faustian bargain" with the nation:

> At NASA's urging, the nation committed to building an amazing, if compromised, vehicle called the space shuttle. When the agency did this, it accepted the bargain to operate and maintain the vehicle in the safest possible way. (97)

How, then, are we to interpret a thesis of the normalization of deviance? Where did it originate? An influential article by Starbuck and

Milliken (1988) argued that "repeated successes and gradual acclimatization influenced the lessons that NASA and Thiokol were extracting from their shared experience" (331). Vaughan cites this article on several occasions. Starbuck and Milliken also argue that "NASA's incremental changes . . . were creeping inexorably toward a conclusive demonstration of some kind," and they supply an explanatory analogy: "A frog dropped into a pot of cold water will remain there calmly while the water is gradually heated to a boil, but a frog dropped into hot water will leap out instantaneously" (337). That is informative about frogs and may apply in a figurative way to human beings in some situations. It is clear, however, that the participants in the teleconference *were not frogs, knew they were not frogs, and knew they were in hot water.* They also knew they were involved in an abnormal if not unique argument. Unfortunately, they chose not to leap out of the hot water.

Starbuck and Milliken also argue that managers and engineers pursue partially inconsistent goals—again implying two different cultures. This would explain the well-known incident when the Executive Vice President at Thiokol asked the Vice President for Engineering to take off his engineer's hat and put on his manager's hat. Engineers stress the decision premise of reliability-safety, while managers stress the *topoi* or premises of cost and schedule or time. In Feynman's organizational communication experiment the engineers said the risk of failure was 1:200 or 300 while their manager gave it the probability of 1:100,000 (repeated by Jensen, 332, and Vaughan, 274). Because of this semantic-information distance between manager and engineers, it must have been *easier for managers than engineers to recommend the decision to launch.*

Messerschmidt (1995) builds on the arguments of Starbuck and Milliken by bringing a feminist perspective to bear on the case because all of those involved in the launch decision were *men.* The stronger "managerial masculinity" overruled the weaker "engineering masculinity" in the Thiokol decision-making process. The conclusion of this article is startling, to say the least:

> In short, the data on the space shuttle *Challenger* explosion provides empirical support . . . [that] one kind of masculinity (manager) was constructed through the commission of corporate crime, while contemporaneously another type of masculinity (engineer) was constructed through resisting that crime. (18)

The language is harsh but there is an element of truth to this explanation—that managers and engineers represented different cultures, an idea I shall deal with in a later chapter.

Vaughan doesn't go quite that far, but in the clearest and most passionate passage of her book she gets close to it while also revealing part of her identification with a subgroup of the principals she interviewed:

> The NASA case follows the classic pattern for organizational misconduct: middle managers were assigned normative responsibility and left "twisting in the wind," while more powerful administrators—some outside the NASA organization, who had acted years earlier in ways that influenced the outcome—were not. (409)

At the explicit level, "misconduct" is not so far from "crime." At the implicit level there is the allusion to the Watergate scandal, in which the person at the top left middle managers twisting in the wind; this calls up memories of high crimes and misdemeanors, particularly when one recalls that the same President—Richard M. Nixon, who left his White House aides twisting in the wind—also forced the original budget cuts that made the space shuttle a difficult project to implement from moment one. More important, Vaughan's expression of sympathy and empathy helps explain why she reverses her early inclinations regarding responsibility for the disaster and becomes motivated to advocate the normalization of deviance as a way of exonerating middle managers: blame the system, not the individual. Vaughan's allusion to "more powerful administrators" outside NASA had to include presidents of the past. This could be her reminder that the White House and Congress may well have influenced the accident by means of decisions to use the shuttle as a kind of space truck without sufficient money to support it. It also makes explicit that Vaughan did hold individuals accountable to some degree.

Similarly, in the final chapter of *No Downlink*, the "*Challenger* Syndrome," Jensen reverses his initial beliefs about blame in the tragedy. I find this chapter to provide a satisfying summary and conclusion. Jensen talks sensibly about large organizations and the inevitable problems of filtering and suppressing information "traveling upward." But that may not be the most important lesson, says Jensen:

Great strength of conviction may be demanded for an individual to speak his or her own sense of what is right and decent. . . . You have to set aside all thought for yourself if you are going to take on the organization's logic and its dissection of reality single-handed. The personal cost can be very high. But often the only ones left to stand guard over a system's social ethics will be a few *responsible individuals* who have suddenly had all they can take. The systems are not going to guard themselves. (370, emphasis added)

I added emphasis to the words *responsible individuals* in order to set up two important points. The first is that Jensen's reminder about the place of the *individual* makes clear how his position differs from Vaughan's more sociological position, summarized by some as "blame the system, not the individuals."

The other reminder that the term *responsible individuals* evokes is the concept of "automatic responsibility" discussed in Chapter Four. That concept was defined by von Braun in the first of his few formal management directives. It fixed responsibility in any individual who had the technical competence to recognize and fix a problem; if the individual was unable to fix it, he or she was to communicate it up the line so that technical strength could be brought to bear on the problem. Every individual was supposed to make NASA a smart buyer and to keep the technical lines of communication open. Automatic responsibility is first and foremost a system of professional and individual accountability.

Both Jensen and Vaughan have a great deal to teach us. I admire much of Vaughan's work and recommend it as a reference volume. The idea of normalization of deviance makes sense much of the time—she may have been inspired by the phrase of Starbuck and Milliken, "gradual acclimatization." The idea helps us understand why NASA could continue to launch the shuttle despite the observed erosion of the O-rings. Jensen, on the other hand, is correct in emphasizing the importance of the responsible individual. The common ground between the two positions would seem to be a communicatively based position in which one looks to both the system and the individual for the best solution to organizational mysteries. My own opinion is that Vaughan is correct in emphasizing the power of the system and structure; Jensen is correct in stressing the individual. Upon reflection, the accident *wouldn't have happened if there had been one individual in management who had said "No, this isn't the way we do it."*

The next chapter will return to the press coverage of the second shuttle accident in order to solve the technical and organizational mysteries. ✦

Chapter Six

The Mysteries of *Columbia* Continue

This book's narrative of the investigation ended after the first week—on Saturday, February 8, 2003, to be exact. This chapter picks up the story again on Sunday, February 9, carrying it to the day that the Report of the Columbia Accident Investigation Board (CAIB) was issued, August 26. There are fewer articles to consider, for one thing, because the Bush Administration's build-up to, and execution of, the War in Iraq crowded other stories out of the news columns.

The search for debris continued apace with the search for historical causes of the accident. David Barstow and Michael Moss contributed a piece on February 9 to the *New York Times* Archives in which they summarized the opinion of a broad range of experts inside and outside NASA in this conclusion: "Years of budget cuts in the shuttle program—cuts that had shed more than 10,000 engineers, technicians and quality control employees—were potentially imperiling the lives of astronauts." Dan Goldin had been the Administrator of NASA during the 1990s; his slogan was "faster, better, cheaper," cutting back NASA employees and increasing dependence on private contractors. Even Goldin thought he'd gone too far, declaring at one point a shuttle crisis and asking for more money.

On Monday, Nick Madigan filed a *New York Times* story with this by-line: Houston, Feb. 9. It said that NASA employees were struggling to cope with the accident, putting on a game face and trying to get back to normal. Some who worked closely with the astronauts needed more emotional help than others, but for most people it was therapeutic to get back to work. A line about a pub near JSC reminded us, if unintentionally, that there were still some mysteries to solve: "At Sherlock's Baker Street Pub and Grill, a tavern near the space center that is frequented by NASA employees and decorated with astronaut memorabilia, business has been down since the disaster."

The Avuncular Anchor

When I went to work for NASA I was advised to watch the *CBS Evening News* because Walter Cronkite, the avuncular anchor, knew more about the space program than any other broadcaster. "He also gives us more coverage," I was told. Cronkite came up in a story filed by John M. Broder on February 10. Mr. Cronkite was described as having narrated on television the most exciting part of the space drama, the Apollo program. Broder interviewed the broadcaster about the *Columbia* accident and was told by Cronkite he was surprised and gratified when the agency opened itself up to the press.

> I've been a little less confident since they started weaseling on the possible cause, going back and forth. . . . My first newsman's hunch was that they decided they'd been so candid that they were beginning to look culpable, and started climbing back on their story.

Broder also reported that Ron Dittemore, the shuttle program manager, would no longer conduct daily briefings. Mr. Dittemore had reversed himself several times as to whether the foam debris could have been a factor in the accident.

Sean O'Keefe's appearance before a joint House-Senate hearing on the *Columbia* accident was featured in an article on February 13, by Richard A. Oppel, Jr. O'Keefe got sharp bipartisan criticism about the Board investigating the accident. The lawmakers persuaded O'Keefe that the Board had insufficient independence and latitude after a four-hour hearing; O'Keefe put NASA to work that night by changing the charter of the Board, loosening its ties to the space agency. Mr. O'Keefe made an analogy in response to a question by Senator Olympia Snow, Republican of Maine, about the foam debris: "The circumstances here were it came off of the external tank as the entire shuttle orbiter system was traveling at 3,600 miles an hour. The piece came off, dropped roughly 40 feet at a rate of something like 50 miles an hour," the equivalent "of a Styrofoam cooler blowing off a pickup truck ahead of you on a highway." *(Everybody knows Styrofoam can't hurt a car, I thought.)*

Other articles in the *Times* reported that one branch of the fault-tree analysis dealt with atmospheric effects. It seems there was a solar storm in space not long before the shuttle returned to the earth's atmosphere. Could this activity have caused damage? The *Times* also published an exchange of e-mail messages between two NASA employees obtained on the NASA website. An engineer, Robert H. Daugherty, who worked at NASA's Langley Research Center in Hampton, Virginia, responded to an e-mail message query from David F. Lechner, a contractor's employee at NASA's Johnson Space Center, about possible damage to the left wheel well.

> I talked to Carlisle a bit ago, and he let me know you guys at M.O.D. [Micrometeoroid and Orbital Debris] were getting into the loop on the tile damage issue. I'm writing this e-mail not really in an official capacity, but since we've worked together so many times, I feel like I can say pretty much anything to you. And before I begin I would offer that I am admittedly erring on the side of absolute worst-case scenarios and I don't really believe things are as bad as I'm getting ready to make them out.
>
> But I certainly believe that to not be ready for a gut-wrenching decision after seeing instrumentation in the wheel well not be there after entry is irresponsible. One of my personal theories is that you should seriously consider the possibility of the gear not deploying at all if there is a substantial breach of the wheel well.
>
> While it is true there are thermal fuses in the wheel, if the rate of heating is high enough, since the tire is such a good insulator, the wheel may degrade in strength enough to let go far below the 1100 p.s.i. [pounds per square inch] or so that the tire normally bursts at. It seems to me that with that much carnage in the wheel well, something could get screwed up enough to prevent deployments and then you are in a world of hurt.

Mr. Daugherty then outlined seven scenarios, such as landing with flat tires, belly landings, and ditching/bailing out. His last paragraph: "Admittedly this is over the top in many ways, but this is a pretty bad time to get surprised and have to make decisions in the last 20 minutes. You can count on us to provide any support you think you need." The publicizing of the e-mail exchange raised a flap. Reporters Schwartz and J. Broder contributed an article the same day the message was printed, February 13, in which they quoted James M. Heflin, Jr., NASA Mission Operations Chief Flight Director, as saying that Mr. Daugherty was merely "what-iffing, which is something we do a lot of."

On Friday, February 14, the Columbia Accident Investigation
·Board (CAIB) announced what seemed to be progress in a statement they released to the press claiming that a hole had been burned into the aluminum skin of the left wing. "Preliminary analysis by a NASA working group this week indicates that the temperature indications seen in *Columbia*'s left wheel well during entry would require the presence of plasma." In an article by Schwartz and Glanz, plasma is defined as a gas "so hot that it glows and conducts electricity. Plasma torches slice through steel as if it were butter, and high technology incinerators use plasmas to break down garbage into its constituent molecules." The reporters concluded that the engineers must have abandoned an earlier theory of the accident— that a damaged or lost tile around the wheel could have allowed heat to be conducted into the well, where sensors first showed a temperature increase. In other words, plasma must have entered the leading edge of the wing, where there are no sensors, and flowed into the wheel well, driving up the temperature reported by the sensors.

> The board did not address the central mystery of how the hole was created. Early suspicions focused on damage from a piece of falling foam insulation that struck the wing about 80 seconds into the launching, and the new finding does not rule out that problem as a possible cause. But since then, other possible sources of damage have also been considered, including collision with space debris or meteoroids.

The central mystery remained. Two weeks after the accident, February 15, William J. Broad of the *Times* wrote that Admiral Gehman, Chairman of CAIB, declined to elaborate on the finding of the hole in the skin of the wing, saying that many theories, including sabotage, were still on the table. But reporter Broad also talked to a professor at Carnegie Mellon, Paul S. Fischbeck, who had studied the thermal tile for NASA and said the hole ruled out certain theories, such as a navigational error, and drew attention to the possibility that debris damaged the tile on the wing. "If you hear hoof-beats, think horses, not zebras," said Dr. Fischbeck.

John Noble Wilford, long-time science writer and Pulitzer Prize winner at the *Times,* wrote an article the same day saying the investigating board might never get to the root cause of the accident. In the case of the *Challenger,* investigators knew two weeks after the accident that the O-rings had caused the accident. And if CAIB couldn't identify the root cause, that would lead to an agonizing decision about whether or when to fly the shuttles again without knowledge of what went wrong with *Columbia*.

LeRoy E. Cain, the flight director of the *Columbia,* gave a news conference on the two-week anniversary of the disaster. Although Mr. Cain mixed metaphors in this answer, his meaning is clear:

> A. O.K. Certainly the possibility of hot gas in the wing is something that is on our fault tree. We're going to basically go wherever the data leads us. And what I mean by that is, we will, now that we have put together a fairly extensive fault tree, the next part of the process that we will endeavor to get into will be to begin systematically eliminate legs on that fault tree. As of yet, we haven't eliminated legs on the fault tree.

A reporter pressed him on whether the flight control team could have done anything to change the outcome of the flight. He replied that "there isn't anything in my estimation that the flight control team could have done differently or should have done differently."

On that same day, Saturday, February 15, Schwartz reported that Admiral Gehman had arrived at the Marshall Space Flight Center in Huntsville, Alabama. The statement that had been released by CAIB on Thursday declaring that plasma entered through a hole in the wing was now considered the single most significant revelation about the accident. Admiral Gehman expressed confidence in unraveling the case. "These accident investigators have solved mysteries with a lot less data than we have," he said. We're only at the "beginning of the beginning." He also said the Board would soon add new members, experts in aerodynamics and thermal engineering, a sign that the Board is operating with independence from NASA. The next day he announced that Sheila E. Widnall, aeronautics professor at the Massachusetts Institute of Technology and former secretary of the Air Force, would be joining CAIB.

"LOSS OF THE SHUTTLE: THE MISSION; After Liftoff, Uncertainty and Guesswork." Houston, Feb. 15. David Barstow wrote this disturbing article in an attempt to reconstruct the events leading to the tragedy. LeRoy E. Cain, 39, *Columbia*'s flight director, said that when he realized the shuttle was breaking up over California he thought of that mysterious piece of debris at liftoff—"That was the first thing that entered my mind."

"While standing by their initial conclusions that the debris posed no threat, Mr. Cain and other senior managers have been humbled enough to acknowledge that some of NASA's best and brightest might have gotten it wrong." They had relied on a computer modeling program run by Boeing called Crater, but there was uncertainty and guesswork about what it meant. Were the flight director and his team too optimistic about the debris strike?

"My only question to the team is how complacent did the team become at looking at these things?" The question came from Carl Meade, a retired astronaut who took a ride on *Columbia* in 1992. "And if complacency ruled the day, that's atrocious. That's a shame." Mr. Meade added, "If you hit a bug going Mach 2, you're going to damage a tile. I don't buy the arguments that say that it was not a big deal."

Officials at Boeing declined to discuss their application of the Crater modeling program. A camera at Cocoa Beach, 15 miles south of the launching pad, was mounted on a telescope that was out of focus. The flight team could have used powerful Air Force telescopes in Hawaii and New Mexico, as well as spy satellites, to take a good look at any potential damage from the debris strike. The option was pointless because Mr. Dittemore had told reporters soon after the accident, "Even if I had information, I can't do anything about it."

A Changed Culture?

A long, devastating article on NASA's organizational culture written by Glanz appeared in the *Times* Archives on February 18. The article begins in a Pentagon office on February 1, 1958. Wernher von Braun, head of the Army's missile research and development, and a brilliant young astronomer named James Van Allen were waiting for word from California that our first satellite had completed its first orbit. "It is in orbit" was the word.

It had taken them only three months to duplicate the Soviet accomplishment of *Sputnik 1*. It set off an unparalleled set of achievements for the United States, landing men on the moon, firing probes on to Venus, Mars and Jupiter. NASA had assembled an impressive talent pool of scientists, engineers, and managers. But by February 18, 2003, a number of scientists, engineers, and historians of technology were telling the reporting team that "NASA has lost its status as a technical powerhouse."

Dr. Van Allen is still active in science 45 years later at the University of Iowa. The Van Allen Radiation Belt around the earth is named after him; he had worked with the space agency to confirm the existence of the belt by means of space flight. "There's been a steady decay in the competence and the feeling that you're not dealing with scientific peers," said Dr. Van Allen. Reliance on private contractors had left personnel at the Goddard Space Flight Center and the Marshall Space Flight Center with "little hands-on expertise."

"They don't really know what's going on," Van Allen said. "They do what they are supposed to do, in a very narrow sense, on a day-to-day basis." Van Allen was echoed by Daniel Baker, director of the University of Colorado's laboratory for atmospheric and space physics. Baker had led a similar lab at Goddard from 1987 to 1994. During the 1990s, when NASA cut some of the last in-house programs that let young engineers build satellites, the decline in talent went from steady to steep.

Howard McCurdy, who outlined the original technical culture we examined in Chapter Three of this book, was quoted in this same article as saying the reduction of force after the Apollo Program amounted to "basically letting their good people go." The NASA budget dropped from nearly six billion dollars in 1966 to 3.3 billion in 1974. He mentioned some findings from a "culture survey" of NASA he had done in 1988. Two of the statements he put to 700 NASA scientists, engineers, and managers on a questionnaire were: "Since I came to work for the agency, NASA has lost much of its in-house technical capability," and "NASA has turned over too much of its basic engineering and science work to contractors." The first statement got agreement from 63 percent of the sample; the second statement got agreement from 78 percent.

In 1995, two major NASA contractors, Boeing and Lockheed Martin, formed a consortium with the name United Space Alliance.

NASA gave United Space Alliance responsibility for day-to-day operations. Glanz interviewed Allan J. McDonald about the outsourcing. McDonald was one of two Thiokol engineers who recommended against the launch of *Challenger* (the other was Roger Boisjoly). He said that while NASA outsourced its operations in the 1990s, the contractors who got the work were themselves losing technical talent in corporate consolidations and cutbacks. He also said managers "didn't have the background and knowledge" to do great engineering.

According to Glanz, there is a decline of talent in the space program as well if one accepts patents and scientific papers as a criterion; NASA's numbers in these areas have been going down for a decade. Young engineers are hard to recruit to the agency. The percentage of workers at NASA younger than 35 has fallen from 28 percent in 1992 to 11 percent. The remaining engineers at NASA and its aerospace contractors are getting older, near retirement age. In addition to these reasons, NASA may have difficulty recruiting the top talent because in comparison with computers and nanotechnology, space hardware is considered an "old" technology.

If it weren't bad enough to have its culture described as declining into incompetence, NASA was facing frequent criticism from scientists. This criticism had been going on since NASA was created in 1958, but after the loss of life it regenerated. Glanz and Oppel surveyed a number of critical scientists in their article on February 24, contrasting their claims with those of NASA's Sean O'Keefe. The scientists said that the financial and human cost of the shuttle program couldn't be justified in the name of scientific progress. Practically all of the scientific experiments on the shuttle could be conducted on robotic flights. O'Keefe dismissed such criticism by saying the International Space Station "is the most extraordinary scientific and research capacity that collectively this many countries could ever have dreamed up."

NASA had powerful defenders against the scientific critics. The House Republican Leader, Tom DeLay, represented a district in Texas in which many NASA employees live. DeLay said there was no option to scale back human space flight, "You won't have the same rewards on investment if you do it by robotics." Recent attempts in Congress to cut back human space flight had been defeated. Senator Bill Nelson, Democrat of Florida, a state where space flight is big business, said, "You can get science on both kinds of missions," noting that it took astronauts to fix the Hubble Telescope. He could have added that it takes astronauts to continue to service the Hubble. NASA was caught between scientists critical of manned spaceflight and those in the astronomical community who rose up in arms when the agency considered saving money by bringing the Hubble down while finishing its replacement, the James Webb Telescope. Astronomers claimed the Hubble had become the most important technique for scientific discovery ever in their field; they admitted to being addicted to the Hubble and said they could not do without it for even a short period of time. O'Keefe's decision in mid-January, 2004, to let Hubble die evoked outrage. The U.S. Senate directed the administration to get a second opinion from the National Academy of Scientists.

On September 21, 2003, the unmanned spacecraft *Galileo* burned up in the atmosphere of the planet Jupiter. The feats of this craft were summarized by the *Economist* magazine (Sept. 20–26, 75–76) under the title "Magnifico!" *Galileo* was launched in 1989, three years after the *Challenger* accident. Its main antenna was supposed to transmit data back to NASA, but it jammed when controllers tried to open it. This should have ended the mission of *Galileo* because its backup antenna could transmit only ten bits of information per second, the equivalent of one letter in our alphabet. But, in what the unidentified writer for the *Economist* would call "one of the most remarkable feats in the history of computer science," the controllers reprogrammed *Galileo's* computer to exploit a modern data-compression method. They then cobbled together a system using a cassette deck designed for other purposes to record data when *Galileo* was cruising by something interesting, such as Jupiter's moons. The data in the tape recorder were transmitted to earth during less interesting periods. NASA had extemporaneously put the bird in almost perfect working order.

Galileo's discoveries were remarkable. It fired a probe at Jupiter to give us details about the planet's air (less oxygen and sulphur but more helium than expected); it observed for the first time that an asteroid, named Ida, had a moon; and the greatest discoveries were related to the four moons of Jupiter first seen by Galileo Galilei, the Italian astronomer of the 1600s. The spacecraft confirmed that the biggest volcanoes in the solar system are on Io, one of four Jovian moons. Another, named Ganymede, the largest moon in the solar system, has an unexpected magnetic field. A third moon, Callisto, may have an ocean but this is not certain. Europa, however, does have water under an icy crust, perhaps more than Earth. If life on Earth began in the ocean, then Europa might have life of some form. For that reason *Galileo* was allowed to burn up in Jupiter's atmosphere on September 21, 2003, rather than risk contaminating Europa. It would be a tragedy on a trip in the future to find, say, *E Coli* on Europa.

The *Economist* said that this was one of America's greatest successes, the kind of thing NASA does best, pushing back the frontiers of knowledge "rather than keeping underemployed astronauts in low Earth orbit, and occasionally killing them" (75). The magazine didn't report that a space shuttle had launched *Galileo* in 1989. And in a related story with the title "Planetophagy," the magazine reported that the Hubble Space Telescope, among others, had recorded images of a star named v838 Monocerotis as it got brighter and then dimmer, changing colors. Two astronomers claim to have solved the mystery: the so-called red giant had swallowed three of its planets! I'm pleased to report that the Earth's Sun is not expected to eat its planets for five billion years. The *Economist*, however, didn't report that the Hubble Telescope was repaired and maintained by astronauts who got there in a space shuttle.

E-Mail Excerpts

The *Times* released excerpts of NASA e-mail on February 27, obtained under the Freedom of Information Act. Someone at NASA identified only as Wayne, whether that be his first name or last, requested assistance from the U.S. Space Command to take high-resolution images of the shuttle's underside. The request was cancelled by Roger Simpson, with apologies that the request hadn't followed the proper channels. One can infer from the message a desire to save the money it would have cost to get the imagery.

> FROM ROGER D. SIMPSON: Thank you for the enthusiastic response to the request for shuttle support yesterday. Your quick response in arranging support was exceptional and we truly appreciate the effort and apologize for any inconvenience the cancellation of the request may have caused. I know that future requests will be met with the same effort.
>
> Let me assure you that, as of yesterday afternoon, the shuttle was in excellent shape, mission objectives were being performed and

that there were no major system problems identified. The request that you received was based on a piece of debris, most likely ice or insulation from the E.T. [external tank], that came off shortly after launch and hit the underside of the vehicle. Even though this is not a common occurrence it is something that has happened before and is not considered a major problem.

The one problem that this has identified is the need for some additional coordination within NASA to assure that when a request is made it is done through the official channels. The NASA/ USSTRAT [U.S. Strategic Command] (USSPACE) M.O.A. [memorandum of understanding between NASA and its contractor] identifies the need for this type of support and that it will be provided by USSTRAT.

Procedures have been long established that identifies the Flight Dynamics Officer (for the shuttle) and the Trajectory Operations Officer (for the International Space Station) . . . to work these issues with the personnel in Cheyenne Mountain [near Colorado Springs, CO]. One of the primary purposes for this chain is to make sure that requests like this one does not slip through the system and spin the community up about potential problems that have not been fully vetted through the proper channels.

Two things that you can help us with is to make sure that future requests of this sort are confirmed through the proper channels. . . . The second request is that no resources are spent unless the request has been confirmed.

Other e-mail messages indicate Ron Dittemore may have been in error when he said on 5 February that everyone agreed that the debris strike didn't involve a "safety of flight" issue. The e-mail excerpts indicate that on January 27, 11 days after the launch, John Kowal, a thermal analysis engineer at JSC, expressed concern that if the tiles near the left wheel well were damaged, hot gases could enter the wing.

FROM JOHN KOWAL: I talked to Ignacio about the analysis he ran. In the case he ran, the large gouge is in the acreage of the door [of the wheel well]. If the gouge were to occur in a location where it passes over the thermal barrier on the perimeter of the door, the statement that there is "no breeching of the thermal and gas seals" would not be valid. I think this point should be clarified; otherwise the note sent out this morning gives a false sense of security.

There was also an exchange between Robert H. Daugherty, an engineer at Langley Research Center, and Carlisle C. Campbell, an expert on mechanical systems. They agreed NASA should get as much information as possible, including an unrehearsed E.V.A., Extra Vehicular Activity, or space walk by an astronaut to assess the damage. Recall also that Ron Dittemore had told reporters that an E.V.A. could cause more damage than the debris strike.

> FROM ROBERT H. DAUGHERTY: I agree completely. Seems to me that the benefit of an E.V.A. to go look at carnage has more pros than cons. Can't imagine that an astronaut (even on a crappy tether arrangement) would cause MORE damage than he is going out to look for!

Mr. Daugherty's concerns elicited e-mail messages from Jeffrey V. Kling and R. Kevin McCluney, flight controllers at JSC, and from William C. Anderson, an employee of the NASA contractor, United Space Alliance.

> FROM JEFFERY V. KLING: If there was hot plasma sneaking into the wheel wells, we would see increases in our landing gear temperatures and likely our tire pressures. If we actually saw our instrumentation in the wheel wells disappear during entry then I suspect that the gear will not deploy anyway because the wires that control the pyros [explosive devices to open the door] and all the hydraulic valves would burn up, too. Ultimately our (MMACS) [maintenance, mechanical, and crew systems] recommendation in that case is going to be to set up for a bailout (assuming the wing doesn't burn off before we can get the crew out). The rest of the cases are great big what-ifs. Any burn-through damage would be discovered well before then.

> FROM WILLIAM C. ANDERSON: First, why are we talking about this one day before the landing and not the day after launch? To quote Paul Dyne, this is another "Burning Rocks Syndrome" (for the new folks, supposedly a week before we went to the moon on *Apollo 11*—yes, we went to the moon—a scientist popped up with a concern about what would happen if the moon rocks beneath the ladder caught fire from the retro rockets). Anyway, if there were evidence on this flight that we were missing tiles/RCC, I might be worried.

Reporters Wald and Broad produced an article interpreting the e-mail excerpts the same day they were published. They interpreted the messages as an "intense debate" over how much damage had been done to *Columbia* and what could be done to save it.

> It is unclear the extent to which the top shuttle managers were aware of the vigorous debate among the technical team, or whether flight controllers had made any contingency plans in the event any of the worst-case failures occurred.

A NASA spokesman in Houston said they were doing an "exercise, a 'what-if.' " He added, "strangely enough, some of the concerns were close to what might have happened." In the exchange of e-mail messages, some of the options are moot if the damage was so severe that the shuttle couldn't fly with control to an altitude where a bailout would have been feasible or could not have come in for a belly landing without wheels. The article also reports that a spokesman in Colorado for the Air Force Space Command said he was unaware of any NASA request but noted that his command had powerful telescopes in Hawaii, New Mexico, and on an island in the Indian Ocean that could see objects in space in great detail. *(That would have allowed NASA to know the extent and location of the damage and given them time to debate in all deliberateness what could be done to save the astronauts.)*

In Washington, D.C., Oppel reported on February 27, that Admiral Gehman of CAIB met separately with Democrats and Republicans on Capitol Hill. The politicians later revealed to the press that Gehman told them he saw a 50–50 chance of pinpointing the cause of the accident; he told them also that the shuttle started shedding debris earlier than originally thought, over the Pacific Ocean; he told them finally it would take between two and six months for his Board, CAIB, to issue a final report.

O'Keefe Gets Grilled

On Friday, February 28, Chang and Oppel reported that lawmakers on the House Science Committee pressed Sean O'Keefe, NASA Administrator, about why he had learned only on Wednesday, two days earlier, about the e-mail messages in which engineers had debated whether foam debris strikes could have caused catastrophic damage. Representative Anthony Weiner, a New York Democrat, asked O'Keefe why he hadn't learned earlier about the "vigorous debate among experts." And he followed up with this question: "Have you fired anyone for not bringing them to your attention sooner?" O'Keefe's answers were not reported, but in response to a question as to whether decisions should be made at the top of the organization, he said "Could be." He also said officials below him decided that the "level of imagery" wouldn't have been high enough to be helpful. Was that a judgment call in error? "We'll find out," said O'Keefe.

Ron Dittemore and Others Are Reassigned

CAIB made news the same day, February 28. Matthew L. Wald revealed that Board Chairman Gehman had requested that NASA reassign several personnel to avoid the appearance of a conflict of interest. Several of the shuttle managers who were responsible for the mission were also involved in NASA's investigation of the accident. The impression was that they were investigating their own decisions and actions. "The leader of the board, Adm. Harold W. Gehman, Jr., told members of Congress privately on Wednesday that Mr. Dittemore was one of several NASA managers he wanted to be returned to their regular duties to ensure the independence of the investigation." Dittemore had made a good impression with the press during the first few days of his press conferences for openness and candor, but this changed as he reversed himself about the likelihood that the foam debris had been a factor in the accident; he also said there had been *consensus among experts* that the debris strike wasn't a flight safety issue; and finally he had also said there wasn't anything NASA could have done had they known the foam had damaged the left wing. His boss, Mr. O'Keefe, didn't agree.

"NASA Chief Disputes Idea That Shuttle Was Hopeless"—that is the headline of an article by Leary in the *New York Times* on March 1, one month after the accident. O'Keefe told reporters,

> I fundamentally, absolutely reject the proposition that there was nothing that could have been done on orbit. . . . There is positively nothing that would have been spared to try to find out what to do to avoid catastrophe.

O'Keefe recalled the *Apollo 13* flight in 1970; while on the way to the moon the spacecraft experienced the explosion of an oxygen tank. He might have mentioned the exciting film, *Apollo 13*, starring Tom Hanks, in which the astronauts had an oxygen shortage and no contingency plan. Engineers on earth made an inventory of materials on board and cobbled together a technical solution to the problem.

The astronauts returned safely to the Earth. He might also have mentioned how NASA creatively and extemporaneously fixed *Galileo* and the Hubble Space Telescope.

O'Keefe acknowledged that many engineering concerns weren't communicated to higher levels of management. At the same time he questioned whether top managers should be involved routinely in deciding technical questions. "Mr. O'Keefe, who is a management expert and not an engineer, said he didn't feel competent to handle complex engineering questions and preferred to let the professionals at lower levels take responsibility for evaluating engineering risks and for making decisions."

On Sunday, March 2, Schwartz named names in discussing what shuttle managers would be reassigned. Letters from CAIB requesting the reassignment of NASA personnel were released without the names, but Schwartz identified Linda Ham and Ralph R. Roe, Jr., leading members of *Columbia*'s mission management team, as people identified by NASA officials. They would no longer participate in the investigation, nor would Ron Dittemore. A spokeswoman for the Board, Laura J. Brown, said Admiral Gehman

> wants to make it clear he's not questioning the integrity of the individuals, or the professionalism of any of the people involved. . . . It's a question of people not investigating themselves, putting people in a position of being in conflict.

On March 5, Wald and Leary reported that four members of CAIB had held a wide-ranging briefing for reporters, saying they had found evidence of damage to the left wing but couldn't yet determine whether it was the incursion of plasma into the wing or the heat of re-entry. Wald and Leary also quoted Sean O'Keefe as suggesting that reporters were overstating the importance of the e-mail messages released the week before. He said it was all speculation by the press until CAIB reached its findings. O'Keefe said he had not seen evidence of malice, complacency, or indifference.

CAIB Adds Three New Members to Inquiry

In an attempt to strengthen its expertise and credibility as a body independent of the government, CAIB announced the names and credentials of three academic experts to be added to the Board. Perhaps the most famous is Dr. Sally Ride, professor of physics at the University of California at San Diego. Ride is a former astronaut who twice flew on the shuttle *Challenger* and served on the Presidential Commission that investigated the first shuttle disaster. Also named were Douglas D. Osheroff, Nobel Laureate in Physics, and John Logsdon, director of the Space Policy Institute at George Washington University. Wald's *New York Times* article on Wednesday, March 5, reported the new appointments and the fact the Board would have an open hearing the following day; they would question Ron D. Dittemore, the shuttle program manager, and others.

Wald reported about the testimony in an article on Friday, the day after the hearings. It turned out to be low key, the witnesses being asked to "affirm" that their testimony was true and complete rather than taking an oath. Mr. Dittemore told the Board he was confident he and other managers had had a good view of safety issues. He also said safety was everything: "If we were fish, it's the ocean we swim in." Others testified that NASA's record-keeping was in disarray and that many of its inspectors were nearing retirement age.

Groupthink?

"NASA's Curse?; 'Groupthink' is 30 Years Old, And Still Going Strong." That is the unusual title of a piece by Schwartz and Wald in the *Times* Archives on March 9. The first sentence:

> At NASA, it really is rocket science, and the decision makers really are rocket scientists. But a body of research that is getting more and more attention points to the ways that smart people working collectively can be dumber than the sum of their brains.

(If the mystery comes to turn on the culture of NASA, it could lead to group dynamics and the concept of groupthink, a term coined by Irving L. Janis, then a professor of social psychology at Yale University.)

Schwartz and Wald supply what they say was Janis' definition of groupthink as "a mode of thinking that people engage in when they are deeply involved in a cohesive in-group, when the members' strivings for unanimity override their motivation to realistically appraise alternative courses of action." Janis found this phenomenon in the Kennedy administration's ill-fated decision to invade Cuba's Bay of Pigs and in the escalation of the Vietnam War. And before his death in 1990, Janis thought it applied to the case of the *Challenger* accident. To avoid groupthink, leadership has to ask penetrating questions and listen hard to what seem to be deviant messages. Indeed, leaders should encourage members to play the devil's advocate, and members should step forward to accept that role without fear of being made fools of for advocating a "burning rocks" theory.

(*Was Robert H. Daugherty merely playing the devil's advocate or did he predict the* Columbia *breakup?*) Daugherty, the NASA engineer whose e-mail messages were examined earlier, had expressed concern about possible damage to the wheel well area. NASA arranged for Daugherty to have a long-distance press conference via telephone from the Langley Research Center in Hampton, Virginia. "The conference call was set up by the space agency, apparently to deflect suggestions that its officials had ignored explicit warnings of disaster," wrote Wong in an article on March 11. Daugherty was quoted in the article as saying it was "frustrating" that the media had been misinterpreting his messages as foretelling the future. "Mr. Daugherty . . . decided to 'play devil's advocate' by looking at what might happen if Boeing's [Crater] analysis was wrong." If so, why didn't his attempt to play the devil's advocate shatter the cohesiveness of the groupthink? I see two possibilities. One is that the shuttle management didn't see Daugherty's e-mail messages. The other is that they saw or heard about it, but their cohesive intensity forced them to discount and ignore it. In either case it is not inconsistent with the idea that the *shuttle management's in-group culture* was insulated from engineering analysis, particularly if it didn't agree with their beliefs.

An anonymous senior NASA official spoke to Wong for a *New York Times* article on March 13. The source said a group of NASA engineers had asked the shuttle program manager, Ron Dittemore, to request the assistance of U.S. spy satellites in determining the extent of debris damage done to *Columbia*. The source said Lambert Austin, an engineer at JSC who was speaking for a group of engineers had asked Dittemore, the shuttle program manager, to request satellite images to help assess the damage, but Mr. Dittemore had turned down the request because he didn't think the imagery would be sharp enough and didn't think it would necessarily help in assessing the damage. Mr. Austin and his group were said to be disappointed because they didn't think Dittemore had enough technical knowledge of imagery to make the decision. "When a group of engineers puts forward a request, they're not doing it for grins and giggles," said the anonymous NASA official.

The images "would have made the decision-making process a lot clearer, and that's the goal," said Professor Paul S. Fischbeck of Carnegie Mellon University, an expert on tiles. "They classified this as not a problem, I think, prematurely." Wong also reported that engineers were studying the two solid rocket boosters to determine whether or not bolt catchers could have failed, causing debris to hit the spacecraft. Bolt catchers are about the size of two large tin cans stacked. When the boosters separate from the external tank the bolts holding them explode; the bolt catchers are designed to catch the bolt fragments.

A senior engineer at JSC, Alan R. Rocha, was reported on March 14 (by Wald and Wong) to have sent an e-mail message to superiors five days after the launch saying that he and other engineers saw "big uncertainties" about the extent of damage to the shuttle. Rocha said NASA should "beg" the Air Force and Defense Department for help in getting imagery of the shuttle. This was thought to be the first internal communication at NASA indicating engineers were worried about the reentry and landing of the spacecraft. A senior investigator with CAIB cited Mr. Rocha's e-mail message as "evidence that top NASA managers might have failed to encourage a healthy exchange of ideas and intimidated those below them in the organization." There are now reports of three different groups of engineers asking for imagery. A Boeing spokesman said that "Linda Ham, the shuttle program integration manager, hearing talk that a request for outside help might be coming, polled only the *mission managers* and none said they were making such a request. She conveyed this to Mr. Dittemore (emphasis added)."

"Amateur Tapes of Shuttle Provide Clues to Breakup," another article by Wald, appeared on March 18. A composite of the images, many of which were provided by amateur photographers in the western United States suggests that *Columbia* began to shed debris earlier than believed, perhaps dropping pieces in the Pacific Ocean, despite a smooth trajectory. On Friday, March 21, Wald reported that a data recorder had been found on a hillside in Hemphill, Texas, two days earlier, on Wednesday, March 19. The recorder, made by Bell & Howell, gathers data from about 800 sensors on the shuttle, encouraging NASA officials to hope they "might have in hand the raw material to clarify the mystery." It will take days of careful work to open up the contents. An astronaut, Captain Robert

L. Behnken of the Air Force, visited Hemphill to thank the searchers who found the recorder.

CAIB Hearing and a Double Meaning

CAIB had hearings on March 26, reported by Wald of the *Times,* focusing on NASA's philosophy of safety. It was the third of a series the Board held. The Board was wondering whether NASA understood the effect of contracting out so much work and whether they thought technical problems through to the end. At that time there were 20 private workers for every NASA worker. (*If so, then how could they penetrate the contractors?*) The debris strikes might well be similar to the O-rings in the *Challenger* disaster. In both cases the problem may have been a "message" that NASA didn't read or hear. "There's no question that NASA looks at the shuttle program through a microscope," said Admiral Gehman, head of the Board, but "NASA managers need to stand back and look at it through a telescope." (*That recommendation has a double meaning, I thought. Is Admiral Gehman suggesting the shuttle managers should have recommended telescopes to take more imagery of the stricken spacecraft?*)

NASA had begun to reconstruct the *Columbia* on the floor of a 40,000 square-foot hangar in Cape Canaveral, Florida. It was like a giant jigsaw puzzle to put together the pieces of debris brought to the hangar from Texas and Louisiana. Wald described the puzzle on March 27, writing that although it looked systematic there was a big gap. There were few pieces of the left wing. That very void may be a clue in the mystery. "I think the hardware is telling us something," said the man in charge of reconstructing the shuttle.

Two More Deaths

A helicopter searching for more pieces of the jigsaw puzzle crashed on March 28, raising the total of people killed in the *Columbia* incident from seven to nine. Charles Krenek, 48, of the Texas Forest Service, and Jules F. Mier, Jr., 56, a pilot with Papillon Grand Canyon Helicopters, were killed when their machine crashed in the Angelina National Forest. The helicopter was one of 28 hired by the U.S. Forest Service to help with the debris search. Information I received from an anonymous source said NASA wanted them to fly low and slow while looking for debris, raising the risk of accidents. (*Will there be a safety investigation of these two deaths?*)

"Space Agency Culture Comes Under Scrutiny" is the headline of an article written by Schwartz and Wald in the March 29 *Times* Archives. CAIB investigators are using story boards to chart the flow of communication in the organization. The cartoonlike panels are used by movie studios and advertisers to visualize who says what to whom; it sounds like a pictorial kind of network analysis used in communication and sociology. They are using the story boards to see where information, i.e., messages, flowed and where they were blocked. There was a flurry of e-mail messages from groups of engineers worried that the left wing of *Columbia* might have been at risk during re-entry:

> But the shuttle program manager, Ron D. Dittemore, and other senior officials including the shuttle program integration manager, Linda Ham, did not follow up on the concerns by requesting that the shuttle be examined for damage by telescopes on the ground or satellites in space.

Mr. Dittemore had told the Gehman Board on March 6 that he had fostered a climate of open communication. Some thought that too optimistic and self-serving.

> Lynda Bottos, who worked at the Kennedy Space Center in Florida for United Space Alliance, the chief contractor for the shuttle program, said that NASA and its contractors continued to have a 'corporate culture of denial' that led them to soft-pedal safety problems instead of reporting them officially. Ms. Bottos was an accident investigator for two and a half years, until late last year.

But CAIB said it wouldn't single out managers in the shuttle program to assign blame for the tragedy; instead, they are taking a deeper look into the NASA culture to see if the agency had reformed itself by favoring the free flow of ideas. NASA had been making efforts to reform its culture since the February 1 accident. Just this week the employee assistance program at JSC publicized a workshop on "Bullying in the Workplace," the notice for which said "Intimidation, harassment and manipulation play a big part in the work environment, and the 'groupthink' mentality often means that the bully gets away with this behavior." I hadn't thought bullying was part of Irving Janis' formulation of the groupthink phenomenon. *(But could this be a tacit admission of bullying and groupthink?)*

Another possible implication of a mistake is the deal struck by NASA with the nation's spy agency. According to a piece by Leary,

March 28, in the *New York Times* Archives, Sean O'Keefe formalized a new accord by a letter sent to the National Imagery and Mapping Agency, calling for them to "use their assets on a routine basis" to take pictures of shuttles in flight. There would no longer be a formal request procedure to request the imagery; no longer would there be the channels that some managers thought hadn't been used properly, a reason given by them to cancel the requests for imagery after the debris strike on *Columbia*.

On March 31, Wald filed a story saying the recovered data recorder did provide "clues" in the attempt to solve the mystery of the shuttle that broke apart. Heating began one minute earlier than had been thought. The sensors were on panels nine and ten of the left wing; they showed a really high temperature that suddenly dropped off to zero, evidence that either the sensors or the wire attached to them had failed.

Bordering on the Irresponsible?

On April 1, the two-month anniversary of the *Columbia* disaster, Schwartz and Leary reported that a senior engineer faulted NASA for not seeking imagery from outside agencies. NASA released via its website two e-mail messages written by Alan R. Rocha, chief engineer of the structural engineering division at JSC. The first memo, believed to be written on January 22, about ten days before the break-up, said in part:

> In my humble opinion, this is the wrong (and bordering on irresponsible) answer from the SSP [space shuttle program] and Orbiter not to request additional imaging help from any outside source. I must emphasize (again) that severe enough damage . . . combined with the heating and resulting damage to the underlying structure at the most critical location . . . could present potentially grave hazards. . . . The engineering team will admit it might not achieve definitive high confidence answers even with additional images, but, without action to request help to clarify the damage visually, we will guarantee it will not. Can we talk to Frank Benz before Friday's MMT [mission management team]? Remember the NASA safety posters everywhere around site stating, "If it's not safe, say so."? Yes, it's that serious.

This may be the most important memo in NASA's history that *wasn't sent!* Rocha didn't answer phone calls the reporters placed with his office and home on April 1. A NASA spokesman said only

Rocha could answer why it wasn't sent. "I can't speak for Rodney [Rocha]," said the spokesman. On January 21, one day earlier, Rocha did send a message to shuttle managers asking, "Can we petition (beg) for outside agency assistance?"

Mr. Rocha reserved a conference room at JSC for the Debris Analysis Team to discuss the possible effects of debris damage. "When the evidence was presented to Linda Ham, the shuttle program integration manager, however, she declined to file a formal request for outside assistance." It was after that decision that Rocha wrote the memorandum addressed to his superior, Paul E. Shack, and more than 12 other engineers. His concerns about plasma entering the wheel well might have been misplaced, for at this stage of the investigation Admiral Gehman was leaning toward a theory that the gases entered a breach in the leading edge of the wing.

Chang elaborated on the new theory in an article written in the *Times* Archives on April 2.

> Investigators examining the loss of the shuttle *Columbia* now theorize that superhot gases rushed into the left wing through a hole in the leading edge. That idea had increased attention on the 22 reinforced carbon-carbon panels in the left wing, some of which might have been damaged when a piece of insulating foam fell off the external fuel tank during lift-off.

The RCC is so called because it is carbon sheets that are layered and immersed in a carbon resin, thus carbon-carbon. After one of its flights, the sister shuttle *Atlantis* was discovered to have a hole measuring one-tenth of an inch by one-seventh of an inch in the layer of silicon applied to the RCC. Below the layer of silicon technicians found a considerably larger hole in the RCC, eaten away by a flow of oxygen.

A Little Bit of an Echo

"I'm hearing a little bit of an echo here," Dr. Sally Ride was quoted as saying in an article by Schwartz and Wald (April 13). She wasn't talking about the acoustics in the hotel ballroom in Houston where members of CAIB were meeting for two days of hearings. She was referring to what she had heard as a member of the Rogers Commission that investigated the explosion of *Challenger*. Problems with a component were ignored after successful flights, an O-ring in one case, a debris strike being an echo in the second. Another expert,

Diane Vaughan, the author of *The Challenger Launch Decision*, discussed in Chapter Five, gave an interview to the reporters in which she detected parallels in early February when she read reports of "longstanding problems with falling foam and the shuttles' fragile insulating tiles." She watched Ron Dittemore in press conferences say that NASA was comfortable with the debris problem. Vaughan didn't think that anyone in either case had done his or her job badly, it was simply a flawed culture.

To Scapegoat or Not

Vaughan, who was scheduled to give testimony to CAIB, said individual responsibility is important. "But she draws a distinction between assigning responsibility and scapegoating." She also said blame on the part of individuals was implied in the Rogers Commission Report; that, said Vaughan, clouded the conclusion that the mistake was in the culture, not individuals. Once Larry Mulloy had left the agency, it was assumed the problem was solved. The reporters talked to Mr. Mulloy, who said he had been made a "scapegoat." The current Board said it was going to avoid scapegoating individuals, a decision that bothers some at NASA. They say that fixing responsibility for mistakes is important. A NASA employee working on the investigation interviewed by Schwartz and Wald quoted Admiral Hyman Rickover, the "father" of the nuclear Navy: "Unless you can point your finger at the man responsible when something goes wrong, then you have never had anyone really responsible."

I reflected at some length on this issue. Kenneth Burke, a longdistance mentor of mine, wrote as wisely as anyone on the dangers of scapegoating, and I long taught that message to my students and anyone else who would listen. We must always be on guard against this phenomenon. Burke's famous review of Hitler's *Mein Kampf*, "The Rhetoric of Hitler's 'Battle,' " written in 1939, before we knew the full horrors of the Nazi Regime, found the book "nauseating" but useful for lessons we could learn from it. He found that Hitler's scapegoating of the Jews was a "curative" unification of Germans by identifying a fictitious devil. We feel better after we project our own weaknesses onto the scapegoat-devil, the enemy, and destroy them both in a gas chamber. Burke, however, did fix responsibility on Hitler for the Holocaust before we knew about it.

CAIB and sociologist Vaughan seem to be heeding Burke's message to avoid scapegoating. On the other hand, Burke declared Hitler to be an enemy of his and an enemy of all Americans. We seem to have here an extreme paradox or antinomy. An antinomy is a Greek concept (*anti* = against; *nomos* = law), a contradiction between two apparent truths, both of which cannot be true. Burke argues against scapegoating but makes Hitler accountable for his rhetorical crimes as well as his physical ones. In the case of *Columbia* we want to say it's OK to blame the culture, the system, not the individual. But at the Nueremburg Trials of Nazi War Crimes, the justices did not accept a defense that "I was ordered to do these crimes"—or that "the culture made me do it." Individuals were held accountable for individual crimes. people do bad, not the culture

We need to make a distinction between scapegoating and the "fall guy" mechanism. Either one is to be avoided for the sake of justice. But scapegoating is often a generalization about a group; the "fall guy" mechanism can falsely select a single person to blame. Burke is not blaming all Germans so much as he is singling out Hitler as a leader who exploited the prejudices of his people. To view it in another way, it would be a mistake to blame *all* NASA personnel for the death of seven astronauts, even though blaming the culture leans in that direction.

I don't for a moment want to give the impression that crimes were committed by NASA officials. I do believe, however, that NASA officials did make mistakes, mistakes that led to the loss of *Challenger* and its crew of seven astronauts. I believe some responsible parties escaped without blame in that case. By April, I was beginning to believe that NASA officials had made mistakes in the case of *Columbia* as well and that it would be difficult for CAIB, despite what they were saying at that moment, not to lay blame on certain individuals and groups within NASA, in addition to blaming the culture. I tentatively concluded, therefore, that there must be a middle ground between avoiding the practice of scapegoating individuals and the need to make individuals accountable.

In a conversation with a friend, Laura Bloss, I explained my need to find a middle way between scapegoating and making individuals accountable. Perhaps, I said, we need to decide on a case-to-case basis, by a casuistry. She nodded.

"What would be important to you?"

"Is what the person did simply one event in a long chain of events or does it have greater causal force than that?"

Her answer set me off on a search for other casuistic criteria by which to decide cases. I came up with eight criteria and present them now and will bring them up again at the end of the book so as to apply them to NASA and its individual members in the *Columbia* case in a way that will allow the reader to decide what stance is appropriate.

1. *Causal force.* Did the action or inaction of the individual have causal force, or did it amount merely to being one act in a long chain of events?

2. *Hierarchy.* What was the person's formal degree of authority and responsibility?

3. *Values of the culture.* Was the individual's act consistent with the ultimate and avowed values of the culture?

4. *Consequences.* Were the consequences of the individual's act or acts trivial or significant?

5. *Justice.* Does the punishment fit the crime? Would the effect of blame be commensurate with the act?

6. *Defense.* Did the person truthfully deny or accept responsibility for the act?

7. *Actor Agency.* Did the person have autonomy, or control over the act or attribute that offends?

8. *Future actions.* Might the person or persons accountable make the same mistake again?

I'm sure there are other criteria for either deciding which opposing principle to follow or what middle position to take, but these eight may suffice to begin stimulating systematic thought about individual cases.

Whether it was scapegoating or fixing responsibility—or something else—Ron D. Dittemore, in a piece by Schwartz on April 20, was said to be leaving NASA for a job in private industry. He had said early on that he was the "accountable individual" for decisions about the safety of the shuttle *Columbia* and its crew, earning praise from me as well as from reporters who thought his candor and heart during the first few days of the tragedy were admirable. Now they were reporting that Anthony Weiner, a Democrat who represented Queens in the House of Representatives, had said that Dittemore's early candor "seemed 'kind of disingenuous' " in light of the revelations of discussions in which officials rejected requests for outside observation or imaging of the shuttle, and that Mr. Dittemore "had, 'to put it mildly, underplayed' the degree to which concerns about the shuttle's condition persisted within the agency."

After an exchange of e-mail messages with Vernon Miller, Professor of Communication at Michigan State University, I agreed to talk with his class of students who were reading my book, *Organizational Communication Imperatives: Lessons of the Space Program.* We agreed that he would call me on April 24, 2003, with speaker phones in his classroom. I would make a few remarks and the students could then ask questions. That morning I picked up a copy of the *New York Times.* In it I found an article by Leary and Wald. They summarized the CAIB testimony of Robert F. Thompson, 77, a veteran of the Apollo Program and a carrier of the original technical culture. He said that foam hitting the Orbiter could be handled by "high school physics." The article summarized his testimony in this way:

> Something had gone wrong with NASA's Problem Reporting and Corrective Action system, which is supposed to track thousands of details and assure that they rise to the appropriate level of attention.

In short, there was a failure of the organizational communication system that had worked so well in the glory days of NASA. I also quoted Sally Ride as hearing a bit of an echo of one shuttle accident in the other. The students asked some excellent questions and I replied to one by saying that

> if you've read my book about NASA you can grasp that this evidence is consistent with the argument in it . . . that the communication systems used with such success during the Apollo Program were not used in the administration of the shuttle program. In addition, lessons learned in the investigation of the *Challenger* were forgotten prior to the accident of *Columbia.*

On April 30, Wald reported that CAIB was preparing to make some tests. They planned to fire pieces of foam insulation used on the ET (external tank) at the leading edge of the shuttle's wing. Admiral Gehman likened the test to hitting a tabletop with a sledge-hammer. "I might not do any damage to the table," he said, "but it might break a leg." By that I assumed he meant that the effects of a blow might show up in parts of a structure other than at the point of impact.

On May 1, three months after the accident, the media discovered that NASA had written an internal report it intended to give to CAIB. Schwartz wrote in the *Times* Archives that CBS News broke the story and NASA confirmed it. The team that prepared the report was led by LeRoy E. Cain, flight director on duty on Saturday, February 1, when *Columbia* broke apart. He said that if one had known about the damage before re-entry, the best option would have been to jettison 15 tons of nonessential objects on board, but this wouldn't have lightened the shuttle's weight enough to reduce the heat on re-entry. Schwartz recalled that Ron D. Dittemore had said "there's nothing we can do" to save the crew. The space agency's Administrator, Sean O'Keefe, later disagreed vehemently with Dittemore. "The new analysis, however, would appear to support the view of Mr. Dittemore, who has announced that he will be leaving NASA after 26 years." *(Wasn't Mr. Cain investigating his own actions?)*

A reporter jumped at a chance to provide some good news, however small in importance it might be. "Peace, in Experiment Form, Survives the Shuttle Disaster," is the headline in a *Times* Archives article, May 9, by reporter Schwartz. A metal case containing an experiment was found among the debris of the shuttle. The experiment was said to be good and viable by a member of NASA's Astrobiology Institute. The experiment, "Growth of Bacterial Biofilm on Surfaces During Spaceflight (Gobbs)" was the idea of two students: Tariq Adwan, a Palestinian student of biology from Bethlehem, and Yuval Landau, a medical student from Tel Aviv, Israel. Both students watched the launch of *Columbia* on January 16.

Leary of the *Times* reported on May 10 that William W. Parsons would replace Ron D. Dittemore as manager of the Space Shuttle Program. Mr. Parsons had served as the Director of the Stennis Space Center near Bay St. Louis, Mississippi. "I came into this thinking that we're going to fly again, and that's what my job is. It is to find out what we need to fix, to fix it, and get back to flight. I think we can fly the shuttle safely."

Wald's article for May 15 reported on a Senate hearing in which both Admiral Gehman and Administrator Sean O'Keefe testified.

Gehman said the safety organization in NASA looks good on paper, but when you look closely at it, "there is not a there there [sic]." The engineers concerned with safety are too few in number and so poorly supported that there is not much they can do. "They're in the room but they're not supported." Mr. O'Keefe admitted the agency hadn't listened adequately to its engineers around the country. The engineers expressed their concerns in e-mail messages; e-mail had democratized communication within the agency, and yet "the results were an unstructured cacophony that NASA had not made sense of." If one defines *cacophony* as discord, dissent, or lack of harmony, it is something organizations should cultivate—that is another meaning of open communication. What is needed, I submit, is a *structured cacophony.* My prime example, of course, would be the Monday Notes. Sometimes technological "progress" in communication makes it more difficult to make sense of the world.

Rescue Try 'Conceivable'

On May 24, Wald and Schwartz produced an article that challenged NASA assumptions about the possibility of saving the astronauts. Admiral Gehman said that although he wouldn't characterize the odds, it was conceivable that NASA could have saved the seven astronauts. The rescue attempt would have been made by getting the *Atlantis* space shuttle ready in a hurry—say, by skipping preflight checks—allowing the rescuers to reach the astronauts before they suffocated. By minimizing their activities, the *Columbia* astronauts would have had enough oxygen and supplies for another two weeks. The shuttles aren't designed to dock with each other, but they could have maneuvered into a position of flying belly-to-belly, then opening cargo doors on both crafts, allowing the seven astronauts in *Columbia* to make an EVA or space walk into the *Atlantis.* "It isn't easy, it's not even highly likely. But it is conceivable," said Admiral Gehman.

The Wisdom of Yogi Berra

The Board gave a briefing to reporters that was discussed in an article by Schwartz published on May 29 in the *Times* Archives. Brigadier General Duane Deal of the Air Force said that NASA had too few eyes watching the shuttles. Inspections once done by NASA employees were now done by contractor personnel.

Board members said they believed that they had identified another incident of foam falling off the area known as the bipod ramp, the spot that shed the foam that struck the *Columbia*. If confirmed, the other incident, in 1985, would be the eighth identified. Until the board began a full-scale investigation, NASA had identified four such incidents, which General Deal said confirmed an adage attributed to Yogi Berra: "You can learn a lot by watching."

I shall move more rapidly over the investigation into the mysteries of *Columbia* during the long hot summer of 2003. The Board conducted tests that confirmed that foam could damage a shuttle wing. Schwartz reported on June 5 that Scott Hubbard, Director of the Ames Research Center and the NASA representative on CAIB, said "Oh my God! This is something. This isn't just a light bounce." Before the tests, many NASA engineers thought the lightweight foam couldn't harm the shuttle; some privately predicted "the foam would bounce off harmlessly, like a Nerf ball. But Mr. Hubbard said the experiment showed that 'people's intuitive sense of physics is sometimes way off.' " (*Everybody knows that foam couldn't hurt a shuttle.*) My semantogenic theory of the foam seems to have traction even with some of the engineers as well as the managers. Benjamin Lee Whorf was correct in asserting that the *name* of a situation or object can influence behavior.

"NASA Hoping to Fly Shuttle After Fixing Some Hardware and Shifting Managers." That is the title of a story by Wald and Schwartz on June 24. Here are the personnel changes:

> Changes in the management of the shuttle program have also been taking place, beginning with the April departure of the program manager, Ron D. Dittemore. He will be replaced by William Parsons, director of the John C. Stennis Space Center. Arthur G. Stephenson, who had been director of the Marshall Space Flight Center for five years, announced in May that he would step down. The center manages the external tank that has been implicated in the *Columbia* accident.

> And this month, Sean O'Keefe, the NASA administrator, moved Gen. Roy D. Bridges from his job as director of the Kennedy Space Center to run NASA's Langley Research Center, replacing the director of that center, Delma C. Freeman, Jr. Engineers at Langley who were consulted during the flight had argued strenuously for closer attention to the possibility that the shuttle might have been mortally wounded at launching. Mr. Freeman will now head

NASA's new Aerospace Transportation Technology Office, which will work to develop future space vehicles.

Despite NASA's claims that people were not forced out, Howard McCurdy said the job changes and shuttle accident were "associated." He said such changes were common after an accident; they reflect a desire for a "fresh start." There is a danger, however, in thinking that by moving people around, you have fixed the safety culture. Blame and punishment tend to have a blinding effect on lawmakers in Washington. They mistakenly think it is the complete solution to a problem.

[handwritten margin note: need more than surface level change]

CAIB continued to have hearings and issue recommendations. They directed NASA to find a way to repair the shuttle in space. A scientist wrote an op-ed piece in the June 29 *Times* Archives. He said he was "culpable" because he had an experiment on *Columbia* that could have been conducted on a robotic flight.

"Shuttle Mystery" was the title of a June 29 piece by Wald. He wrote that in five months of investigation, CAIB had had perhaps only one "eureka" moment. That occurred in early May when investigators realized "from the wreckage that hot gases flowing through a wheel well that were suspected on the day of the disaster because radio transmissions from the shuttle indicated high temperatures in the landing gear were flowing out, not in." That meant the plasma was coming into the craft somewhere else, perhaps the wing, *not* near the landing gear; the plasma was flowing out of the wheel well.

Five months after *Columbia's* demise, on July 1, a bit of dark humor popped up. Wald and Schwartz reported that NASA's release of transcripts representing internal conversations on February 1 showed that at first the re-entry was so normal that engineers joked among themselves. They laughed about a minor problem concerning "spatial" differences between balloon data and aircraft data. An engineer on a support team working in a nearby room, apparently mimicking the "Church Lady," Dana Carvey's character on *Saturday Night Live*, said, "Well, isn't that 'spatial'?" The mood quickly changed when sensors began to fail; an engineer plaintively asked, "What in the world?"

"NASA Goes Shopping for a Shuttle Successor, Off the Rack," appeared on July 1, over an article by Leary. The agency wanted to build an OSP, an orbital space plane that could reach the International Space Station more cheaply and more safely than the shuttle. They proposed to use existing technology. Sean O'Keefe said they're taking the "Kiss and tell" approach, "Keep It Simple, Stupid, and tell us the best way to get what we need." *(It is likely to be more difficult than that.)*

NASA also announced more changes in the management of the shuttle program, reported July 3 by Schwartz and Leary. The big surprise was that Ralph Roe, the Orbiter project manager, was named to a new job providing independent evaluations of safety and engineering. Roe was one of two NASA managers Admiral Gehman asked NASA to remove from the investigation (the other was Linda Ham). Roe was to help Bill Parsons, who replaced Ron Dittemore as shuttle program manager. Howard McCurdy was asked about the move to put Roe, part of the shuttle management team that had been criticized, into the new position. He said this was an instance of the old technical culture. "He noted that the rocket scientist Wernher von Braun allowed his engineers to make one mistake. 'Von Braun figured that once you made the mistake you'd never make it again.' " Sean O'Keefe was reported in this article to have warned NASA employees at JSC in Houston that CAIB's report would be "really ugly."

Katie Zezima reported on July 10 that Daniel S. Goldin, former NASA Administrator, had accepted the presidency of Boston University. It wasn't announced whether Goldin would ask the faculty, staff, and students to work "faster, better, cheaper."

Linda Ham Meets the Press

Although there had been news reports that Linda Ham had been removed from the investigation of the *Columbia* accident, she didn't speak to the press until July 23. Her first public statement was presented in the *Times* that same day by Wald and Schwartz. She became "tearful" when she described her experience as chair of the mission management team. She said, "I was never alerted to the concerns that were expressed by the engineers working the issue, neither the severity, the potential severity some of them felt about the damage, nor the fact that they wanted the on-orbit image."

She went on to say she and the others were basing their decisions on the best information they had at the time. "Ultimately, she said, 'I don't believe anyone is at fault for this.' " Others disagreed with that assessment. Wald and Schwartz got access to transcripts of management meetings NASA released at the same time Ham made her statement. They wrote that in a crucial January 24 meeting,

> Ms. Ham cut off a NASA engineering manager, Don L. McCormack, Jr., while he was presenting uncertainties and unknown risks from the piece of foam that struck the shuttle some 80 seconds into the flight. Ms. Ham repeatedly stressed in the following discussion that the foam posed "no safety of flight" concern and "no issue for this mission."

The new documents showed a management team that was not on guard and not following the rules. Despite a NASA rule requiring management meetings every day during a flight, Ham's group had held only five meetings during the 16-day flight, also taking a long weekend break during the Martin Luther King, Jr., holiday.

There is some confusion about who knew what. Ham claimed she didn't know "officially" that groups of engineers had concerns about the debris strike. She said she'd "heard informally" about a request for images. She said she spent the better part of a day trying to find out who wanted the pictures. In her briefing on July 23, Ham said top managers had to rely on others. "I don't have the engineering expertise, nor do I have the tools, to do that kind of analysis."

> But to a NASA engineer who works closely with the mission management team, Ms. Ham's arguments sounded like an effort to plead ignorance as a defense, which he called "unbelievable." He said, "It's your job to know the people to ask the questions. Part of it is recognizing your limitations."

By saying that she didn't know about the concerns, Ham was saying we should blame the system, not the individual. By saying she thinks no one was at fault, that it could only be the system, she wanted no individual scapegoating, no fall guys, no personal accountability at all.

The press had by late July turned its attention to Iraq and the Roadmap for Peace in the Middle East. Coverage of NASA mainly dealt with the CAIB Report, which was due to be released on August 26, 2003. The Board had been relatively open with the press, but some mystery remained. Could the Board say with any degree of certainty what happened to *Columbia*? And in addition, what would CAIB have to say, if anything, about communication and culture in NASA?

Schwartz got some clues about the answer to the second question in interviews with Admiral Gehman and sociologist Diane Vaughan; he reported the results in his *Times* Archives article of August 22, just four days before the Report was released. Gehman said that when he began looking into the accident he thought about how to take a broader view than is used in most investigations. In the middle of February Vaughan's publisher, The University of Chicago Press, sent a copy of her book to Admiral Gehman. He read it and urged others on the Board to read it. He invited her to testify at a hearing "and then hired her to help shape the final product. She ended up writing Chapter 8."

Vaughan told the reporter: "This report is truly an intellectual breakthrough in terms of accident investigations." ✦

Chapter Seven

Reading the CAIB Report: Echoes of *Challenger* and a Cultural Fence

Within two days after the *Columbia* accident, NASA's—take a breath before saying the title: International Space Station and Space Shuttle Mishap Interagency Investigation Board—was transformed by Administrator Sean O'Keefe into the Columbia Accident Investigation Board (CAIB), with Admiral Harold W. Gehman, Jr., United States Navy (retired) as its Chair. G. Scott Hubbard, Director of NASA's Ames Research Center was the representative of the agency's field centers. The other six members represented other government agencies such as the Air Force, Federal Aviation Administration, the Navy, and others. Because of the outcry that the government was investigating itself, other representatives were appointed, including people from business, the academic world, and, notably, Dr. Sally Ride, professor of physics, astronaut, and member of the Rogers Commission that had investigated the *Challenger* accident.

Within seven months the Board issued its report. The CAIB Report came out on Tuesday, August 26, seven months after the accident. I got a copy the next day. I regarded it as the last chapter in the police procedural, the detective story, the place where we find the resolution of the mysteries. The report is 248 pages long including appendices. It contains a page "IN MEMORIAM," which lists the names of nine people killed, seven astronauts and the two people killed in the debris search, Jules F. Mier, Jr., Debris Search Pilot, and Charles Krenek, Debris Search Aviation Specialist. The Table of Contents shows a Board Statement, an Executive Summary, and a Report Synopsis. Part One contains the first four chapters: 1. "The Evolution of the Space Shuttle Program"; 2. "*Columbia*'s Final Flight"; 3. "Accident Analysis"; and 4. "Other Factors Considered." Part Two contains chapters 5 through 8: 5. "Why the Accident Occurred"; 6. "Decision Making at NASA"; 7. "The Accident's Organizational Causes"; and 8. "History as Cause: *Columbia* and *Challenger*." Part Three, "A Look Ahead," has three chapters: 9. "Implications for the Future of Human Space Flight"; 10. "Other Significant Observations"; and 11. "Recommendations." Part Four contains three appendices summarizing the investigation, brief biographies of Board members, and the Board Staff.

As I pored over the report to understand *what* the Board was presenting to us, I was also interested in *how* they reached their conclusions. A subheading on page 85 of Chapter 4, "Other Factors Considered," jumped out at me: "4.2 Fault Tree."

As explained earlier, NASA (and its investigating team) used fault trees, graphical representations of all conceivable sequences of events that could have led to the breakup of *Challenger*. All potential chains of causation of the failure were diagrammed. Let's look at what branches got eliminated, or "closed," and which ones remained "open." There were two main possibilities to explain the aerodynamic breakup of the shuttle: (1) the shuttle sustained structural damage that undermined the attitude or angle of re-entry; or (2) the shuttle maneuvered to an altitude for which it was not designed. The first explanation involves damage to the craft at any time before re-entry; the second involves improper trajectory control by the Flight Control System. Data strongly indicated that improper maneuvering was not the problem, dictating that most of the fault tree analysis be concentrated on structural damage.

Fault Tree Analysis

I climbed down through the following branches.

1. Solid Rocket Booster Bolt Catchers

Each solid rocket booster is connected to the external tank by four bolts, one of which weighs approximately 68 pounds. Two minutes into the flight, pyrotechnic charges break the bolts so that the boosters can separate from the external tank and drop into the ocean. The bolts cannot be allowed to strike the Orbiter so there are bolt catchers on the external tank. An unknown metal object was seen separating from the "stack"—the orbiter, external tank, and boosters—during six space shuttle launches. The Board ran tests and found that the bolt catchers didn't measure up to specifications. They eliminated bolts as a cause of the accident, however, because they were so big that their impact would probably have registered on the shuttle stack's sensors. They did recommend, however, that the bolt catchers be tested and "qualified" for future flights.

2. Kapton Wiring

An avid reader of *Superman* comics as a boy, my long-term memory bank must have been the cause of my misreading of the name of this space-age wiring. An early reader of the manuscript correctly wrote in the margin: "Krypton?" Because of previous problems with electrical wiring, it was considered an important branch in the analysis. The Board found no evidence that this was the cause but recommended a new state-of-the art inspection method for the future.

3. Hypergolic Fuel Spill

Hypergolic fuels are those that ignite spontaneously when mixed together. During fueling some of the fuel had spilled on the left trailing edge of the orbiter's left inboard aileron. Two tiles were removed and inspected before lift-off, and the Board ruled the spill was not a factor.

4. Space Weather

Space weather refers to highly energetic particles in the outer layers of the earth's atmosphere. A solar flare occurred just before the shuttle started its re-entry. Measurements of the flare allowed the Board to rule out this factor.

5. Asymmetric Boundary Layer Transition

This branch has a mouthful of words for a title and yet it means a rough surface on a wing, a possible cause of turbulence. The Board pruned this branch by noting that measurements showed *Colum-*

bia's wing roughness was well below the average of the fleet of four shuttles.

6. Training and On-Orbit Performance

Seven flight controllers on this mission were without current certification at the time of the flight. They made a couple of mistakes during the flight; the errors were noticed by the astronauts and immediately corrected. The Board ruled that these matters were not factors in the accident.

7. Payloads

The Board examined all sources of data about payloads from inspections, debris analysis, and other evidence; they found no reason to conclude they were a factor in the accident.

8. Willful Damage and Security

Concerns about terrorism and sabotage were reported in the media from the first moment. The Board visited all relevant space centers to check security systems and used other data to reject this as a factor in the accident.

9. Micrometeoroids and Orbital Debris Risks

The finding of the Board on these issues: "There is little evidence that *Columbia* encountered either micrometeoroids or orbital debris on this flight" (94). The Board did recommend, however, that NASA raise the standards for safety in these areas and change the criteria from guidelines to requirements. Perhaps many Americans might find shuttle flights more interesting if they knew that the ship, whenever possible, flew backwards, or tail first, to prevent debris from hitting the windshield or the leading edges of the two wings.

10. Foreign Object Damage Prevention

Workers sometimes leave their tools and other objects in the Orbiter when working on it. Anything, no matter how innocent it sounds, anything out of place in a rocket can have catastrophic effects. Von Braun used to show to important visitors at Marshall an oily rag a contractor's worker left in a rocket. In another inspection a lunchbox and tools were found. Recent daily debris checks have revealed a success rate of between 70 and 86 percent. The Board didn't consider this a causal factor but recommended that KSC and the contractor be required to comply with industry-standard definitions.

Sabotage was ruled out, as were bolt catchers, flight control errors, payloads, and other hypotheses. What, then, was the answer to the mystery?

Perhaps the most vivid and convincing piece of evidence in the report is a photograph on page 72 taken from above the reconstructed pieces of debris from the *Columbia*. (NASA recovered some 84,000 debris pieces, or about 38 percent of the Orbiter.) The Board said the analysis of the debris revealed

> unique features and convincing evidence that the damage to the left wing differed significantly from damage to the right, and that significant differences existed in pieces from various areas of the left wing. While a substantial amount of upper and lower right wing structure was recovered, comparatively little of the upper and lower left wing structure was recovered. (73)

Tiles recovered in the debris search also offered valuable clues. Tiles from the lower left wing sustained extreme heat damage and more signs of erosion than did any of the other tiles found. This heat erosion was "likely caused by an *outflow* of superheated air and molten material from behind RCC panel 8 through a U-shaped design gap in the panel" (73). Chemical analysis showed that tiles from near that area were covered with molten Inconel, a material found in fittings on the leading edge of the wing. Most right wing tiles were simply broken off because of aerodynamic forces, but the tiles on the left wing show

> significant evidence of backside heating of the wing skin and failure of the adhesive that held the tiles on the wing. This pattern of failure suggests that heat penetrated the left wing cavity and then heated the aluminum skin *from the inside out*. As the aluminum skin was heated, the strength of the tile bond degraded, and tiles separated from the Orbiter. (74, emphasis in the original)

"From the inside out," that phrase seems to explain early puzzles about how sensors could show an increase in temperature in areas away from the likely point of the initial damage—e.g., the wheel well. The rest of the evidence definitely pointed toward the left wing. There was more. Pieces from the left wing RCC panels 9 through 22 had fallen to earth farther west than pieces from other sections. Pieces from panels 1 through 7 were farther east and pieces from left wing panel 8 were found throughout the field of debris, suggesting that the left wing likely failed "in the vicinity of RCC panel 8" (75). The tiles found the farthest to the west came from the area behind left wing panels 8 and 9. Although this evidence, says the Report, is not conclusive of a breach in panels 8 and 9, the pattern does suggest unique damage there. Chemical and X-ray analyses of the RCC panel 8 are also reviewed, leading the Board to find that "Multiple indications from the debris analysis establish the point of heat intrusion as RCC panel 8-left" (78).

A test was set up in which a piece of foam was fired at panels of a flight-worthy left wing. The impact tests were done at the Southwest Research Institute facility. A nitrogen-gas gun used to test bird strikes on airplane fuselages was adapted for the test. RCC panels were the targets. A 1.67-pound foam projectile hit panel 8 at 777 feet per second or about 500 miles per hour: "The impact created a hole roughly 16 inches by 17 inches which was within the range consistent with all the findings of the investigation. Additionally, cracks in the panel ranged up to 11 inches in length" (83).

Finding F3.8-7: "The bipod ramp foam debris critically damaged the leading edge of *Columbia's* left wing" (83). A fuller explanation was presented in the Executive Summary:

> The physical cause of the loss of *Columbia* and its crew was a breach in the Thermal Protection System on the leading edge of the left wing, caused by a piece of insulating foam which separated from the left bipod ramp section of the External Tank at 81.7 seconds after launch, and struck the wing in the vicinity of the lower half of Reinforced Carbon-Carbon panel number 8. During re-entry this breach in the Thermal Protection System allowed superheated air to penetrate through the leading edge insulation and progressively melt the aluminum structure of the left wing, resulting in a weakening of the structure until increasing aerodynamic forces caused loss of control, failure of the wing, and breakup of the Orbiter. This breakup occurred in a flight regime in which, given the current design of the Orbiter, there was no possibility for the crew to survive. (9)

The technical mystery was solved. Significantly, it was the reinforced carbon-carbon that was damaged, despite the early speculation it was the tile. For all the speculation about tile on the wheel well and tile on the wing, it was the RCC panel number eight. It makes one wonder what telescope and spy satellite imagery might have told Rodney Rocha and other concerned engineers. On page 60 the Report contains a shaded box with this title: "The Orbiter 'Ran Into' the Foam." A question is posed: "How could a lightweight piece of foam travel so fast and hit the wing at 545 miles per hour?" The answer is, *it didn't*:

> Just prior to separating from the External Tank, the foam was traveling with the shuttle stack at about 1,568 mph (2,300 feet per second). Visual evidence shows that the foam debris impacted the wing approximately 0.161 seconds after separating from the External Tank. In that time, the velocity of the foam debris slowed from 1,568 mph to about 1,022 mph (1,500 feet per second). Therefore, the Orbiter hit the foam with a relative velocity of about 545 mph (800 feet per second). In essence, the foam debris slowed down and the Orbiter did not, so the Orbiter ran into the foam. The foam slowed down rapidly because such low-density objects have low ballistic coefficients, which means their speed rapidly decreases when they lose their means of propulsion. (60)

(This helps explain to those of us, including engineers and managers, whose intuitive sense of physics is faulty, how a piece of foam could destroy a space shuttle.)

The Mystery of NASA's Culture

There was still some suspense from the other mystery—what would the Board find and recommend about the organization and its management? Had they been contributing factors in the loss of the shuttle and the seven astronauts? I anticipated they would be found as such because Sean O'Keefe, the NASA Administrator, was widely reported to have warned NASA employees that the report would be "very ugly." Part of the answer came in the second paragraph of the report's one-page Executive Summary: "The Board recognized early on that the accident was probably not an anomalous, random event, but rather likely rooted to some degree in NASA's history and the human flight program's culture" (9). But even so, what diagnosis would the Board use?

As reported in the *New York Times* interview, Diane Vaughan's publisher had sent a copy of her book (discussed in Chapter Five), to the Board Chair, Admiral Harold W. Gehman, Jr. He read it and recommended it to other members of the Board. The article said she had served as a consultant to the Board and wrote the eighth chapter of the CAIB Report, leading me to believe that her idea of "normalized deviance," addressed in Chapter Five of this book, might be part of the diagnosis. She had developed that idea in her book on *Challenger*, and as we have seen, she rejected the conventional explanation by the Rogers Commission, by Jensen, by McCurdy, by me, and by others, which stressed problems in communication, including the reversal of the burden of proof.

In its Executive Summary, the Board hints that it will embrace both types of explanations:

> Cultural traits and organizational practices were allowed to develop, including: reliance on past success as a substitute for sound engineering practices (such as testing to understand why systems were not performing in accordance with requirements); organizational barriers that prevented effective communication of critical safety information and stifled professional differences of opinion; lack of integrated management across program elements; and the evolution of an informal chain of command and decision-making processes that operated outside the organization's rules. (9)

In Chapter 6 of its Report, "Decision Making at NASA," the Board concentrates on the lack of appreciation the agency's engineers and managers had for Hazard Reports about the shedding of foam. They accepted events that were not supposed to happen; the Board made this relevant observation: "The acceptance of events that are not supposed to happen has been described by sociologist Diane Vaughan as the 'normalization of deviance' " (130). But they also find that a passage from Richard Feynman's minority report to the Rogers Commission, discussing the O-rings in the case of the *Challenger*, fits the *Columbia* case as well. The report says the "parallels are striking" in introducing a long quotation from Feynman. Erosions and blow-by in the O-rings, giant washers around the joints in the solid rocket boosters, are not what the design expected, said Feynman; they are, rather, a kind of warning that something is wrong. With the pressure of a launch schedule that must be met, engineers can't keep up with conservative criteria, and so, said Feynman, "subtly, and often with *apparently* logical arguments, the criteria are altered so that flights may still be certified in time" (130, emphasis added).

Thus, Feynman had his unique expression of the normalization of deviance theory that included communication concepts. Feynman also believed there was a serious communication problem between NASA engineers and managers, as can be recalled in the discussion of his communication experiment at MSFC in Chapter Five of this book. I drew on Feynman's observation in my analysis of communication problems in the *Challenger* accident (1993) discussed above. Feynman's writings about the first accident prove that the normalization of deviance and miscommunication are *not* incompatible explanations. In fact, I shall argue that the former presupposes the latter in that "subtly, and often with apparently logical arguments," managers accept something that is wrong, not what the design expected.

We need to look more closely at Feynman's version of the idea. He is implying that there are several steps to the process. The first is the recognition of an anomaly, something not expected by the design; there can't be an anomaly unless and until there is discourse about it. Next come the argument or arguments as to whether the level of risk is acceptable or not. The third step is the decision stage, or how arguments about it are received, whether accepted or challenged or changed. For Feynman the drift into acceptance of what the design did not expect is *subtle but not done in silence*; the pressure is provided by schedule demands that link each anticipated flight and by what Feynman called an *"apparently* logical argument."

Testing Feynman's Hypothesis

I shall "test" this hypothesis by seeing how consistent it is with the Board's report. Chapter 6 in the CAIB Report, "Decision Making at NASA," establishes that the new Administrator, Sean O'Keefe, committed the agency to a February 19, 2004, launch date that would lift Node 2, the skeleton of a room addition, to the International Space Station. Engineers felt the date was arbitrary, designed to satisfy the White House and Congress, to prove NASA could meet a deadline and hold down costs. It affected how they approached every launch scheduled before it. Slips, or delays, in the *Columbia* flight STS-107 (January 16, 2003) could affect subsequent launches and thus delay the Node 2 launch. NASA even distributed a computer screen saver to employees with a clock ticking down to February 19, 2004. The Board concluded:

> The agency's commitment to hold firm to a February 19, 2004, launch date for Node 2 influenced many of the decisions in the

months leading up to the launch of STS-107, and may well have subtly influenced the way managers handled the STS-112 [October 7, 2002] foam strike and *Columbia*'s as well. (139)

Note the word "subtly," a variant of Feynman's very word to describe how time pressure could predispose one to accept a faulty argument instead of sound arguments supported by engineering analysis and testing.

The hypothesis that time pressure subtly affected decisions is tentatively confirmed; now we shall examine how technical concerns come into existence and are dealt with. After the foam debris shedding was noticed, the piece was reported to be about the same size as one that had hit the previous flight, STS-112 (October 7, 2002), thus establishing it as close to being in the experience base, almost close enough to being "in-family," a NASA expression meaning "we know it well." (It seems similar to Ludwig Wittgenstein's category of "family resemblance," and again could be a problem of "meaning" in which NASA engineers and managers could be misled into thinking they knew more about the problem than they actually did.)

Nonetheless, two groups would go to work on the debris problem soon. On the day after the launch a group called the Intercenter Photo Working Group looked at the images of the launch. After they got some higher resolution images, they noticed the debris hit the Orbiter 81.9 seconds after the launch. The size of the object created concerns within the Working Group. They requested that the Department of Defense take high-resolution, in-orbit images of the shuttle so that analysts could get the best look at the spot where the piece of debris hit. They shared their video clip and report with the Mission Management Team, the Mission Evaluation Room, and engineers at United Space Alliance and Boeing, two contractor organizations.

NASA engineers went to work on an assessment of potential impact damage, forming a Debris Assessment Team to conduct a formal review. The Debris Assessment Team (DAT) had its first formal meeting five days into the mission, January 21. The meeting ended with an agreement that the highest-ranking engineer in the group would make a request on behalf of DAT for better images; he would go up through the channels of KSC with the expectation it would reach Space Shuttle Program managers. "Debris Assessment Team members subsequently learned that these managers declined to image *Columbia*" (38).

Without images to work with, the Debris Assessment Team was limited to using a mathematical modeling program called Crater, even though it hadn't been designed for this use. They came to the tentative conclusion that some localized heating damage could occur during re-entry, but they weren't certain whether structural damage would occur. Eight days after the launch DAT reported their findings to the Mission Evaluation Room, whose manager gave an oral summary, without data, to the Mission Management Team the same day. The Mission Management Team categorized the debris strike as a "turnaround" issue, a matter of mere maintenance that could be done before *Columbia*'s next flight, and thus did not pursue a request for imagery.

The Board would count three requests for imagery that would be denied, the first being made in person on Flight Day 2 by the Intercenter Photo Working Group; the second made by phone on Flight Day 6 by an employee of United Space Alliance; and the third by a co-chair of DAT on Flight Day 6 by e-mail. They also counted eight "missed opportunities" for NASA managers to understand the nature of the damage before re-entry, including a spacewalk inspection by the astronauts. These missed opportunities would become magnified when the Board later determined that there was a possibility that the shuttle *Atlantis* could have *rescued the seven astronauts on* Columbia *had NASA known about the damage to the left wing*.

How can we explain this fatal series of mistakes? Notice that the DAT engineers were concerned about this deviant, anomalous event and wanted more data, images, to continue their analysis, but it was increasingly difficult for them to have their "concerns heard" by those in a decision-making position. The first stage of Feynman's communicative hypothesis is confirmed. Two groups notice the strike, discuss it, and label it a concern. The second step of the Feynman hypothesis is the argument about the level of risk. The Debris Assessment Team wants more imagery and makes requests for it. The Board clearly introduces the concept of presumption and the burden of proof in these remarks:

> In the face of Mission managers' low level of concern and desire to get on with the mission, Debris Assessment Team members had to prove unequivocally that a safety-of-flight issue existed before Shuttle Program management would move to obtain images of the

left wing. The engineers found themselves in the unusual position of having to prove that the situation was *unsafe*—a reversal of the usual requirement to prove that a situation is *safe*. (169, emphasis in original)

The second step does take place but in an inverted manner. The Debris Assessment Team wants more data, more imagery, so that they will know what to argue about the level of risk. They find themselves needing to prove an argument before they have the evidence to back it up. In addition, the Shuttle Program management would later claim they didn't know about the requests. But it is clear that they had decided the foam debris strike was acceptable, for arguments that were not even *apparently logical* to them. They would complete the third step of the Feynman hypothesis, the decision step, by subtly accepting the foam strike because it was part of the rationale in their flight readiness review. Two of the members of the management team—Linda Ham and Ron Dittemore—believed the rationale for establishing the debris strike as not a safety concern to be "lousy" during their review. Nonetheless, the three steps of the process were present, if in pathological form, proving that the process of the normalization of deviance is constituted in and by discourse or communication.

Other Communication Problems

The Board produced a heading on page 169, "A Lack of Clear Communication," in which they submit that communication did not flow effectively "up to or down from Program managers." After the accident the Program managers claimed publicly and privately that they hadn't known about the engineers' requests and that engineers have a responsibility to communicate their concerns to management. The Board felt otherwise, "Managers did not seem to understand that as leaders they had a corresponding and perhaps greater obligation to create viable routes for the engineering community to express their views and receive information" (169).

Mission Management Teams were found to be breaking NASA's rule requiring them to meet every day during a flight; they met only five times during the 16-day mission. They began the mission with the belief that a foam strike was not a safety issue, conditioned by their own Flight Readiness Review in which previous strikes hadn't produced failed flights. The engineers didn't share that belief, became concerned about the foam strike, made highly accurate calculations about the size of this particular chunk of debris, and wanted more data in the form of images of the shuttle. Managers were more concerned about the schedule and the proper channels of communication than they were about safety. Thus, Finding F6.3-15:

> There were lapses in leadership and communication that made it difficult for engineers to raise concerns or understand decisions. Management failed to actively engage in the analysis of potential damage caused by the foam strike. (171)

The Board also produced a finding that the engineers were improperly asked to prove the vehicle was unsafe rather than prove it was safe. The Board would later demonstrate the danger of reversing the burden of proof: "Imagine the difference if any shuttle manager had simply asked: 'Prove to me that the *Columbia* has *not* been harmed' " (192).

Organizational Causes

In its Chapter 7, "The Accident's Organizational Causes," the Board came up with its own definition of culture:

> Organizational culture refers to the values, norms, beliefs, and practices that govern how an institution functions. At the most basic level, organizational culture defines the assumptions that employees make as they carry out their work. It is a powerful force that can persist through reorganizations and the reassignment of key personnel. (177)

Although I prefer the definitions with which we began Chapter Three of this book because of the clause, "communicatively structured, historically based system," and the emphasis on language and *meaning*, I can live with the Board's definition if we assume communication is included in their phrase, "practices that govern how an institution functions."

The Board looks at history in making its cultural analysis, correctly examining the kind of leadership NASA had during its four decades. It looks closely at the leadership of Daniel S. Goldin, an executive of the aerospace contractor, TRW, Inc., who was nominated on March 11, 1992, by the first President Bush to replace Richard H. Truly as Administrator of NASA. He would serve almost a decade in that position. An article by Broad in the *New York Times* the day after his nomination in 1992 reported that Goldin was expected to be more receptive to White House views than was his predecessor. An anonymous White House source was quoted as saying of Goldin, "He's obviously outside the NASA culture" (Broad 1992, A1).

Goldin was also reportedly expected to develop tighter bonds between NASA and the private sector. He is, or was at one time, an expert in space robotics; at TRW he was involved in such projects as Brilliant Pebbles and Brilliant Eyes, space satellites that were portions of the Strategic Defense Initiative ("Star Wars"). I watched Goldin in a television interview with Robin McNeil of PBS after astronauts successfully reorbited an errant satellite in May of 1992. He seemed knowledgeable, at ease, and quietly impressive as he assured his viewers that NASA would seek a "balance" between "human and robotic" projects. The press also reported that Goldin was keen to achieve von Braun's brightest dream, a trip to Mars and back by U.S. astronauts.

A profile of Goldin and his activities between April and August was published by the *Los Angeles Times* on August 5 (Pasternak 1992). During that four-month period, Goldin had shaken up NASA. He began soliciting opinions from the 24,000 employees of the agency, asking why the United States needs a space program. Goldin said he intended to hold town hall meetings on the subject all across the country.

Fasterbettercheaper

In frequent pep talks, Goldin told NASA employees they were going to have to do more with less money: "Fasterbettercheaper" came out of his mouth as one word (Pasternak 1992, A12). He had set "red" teams against "blue" teams in order to come up with cost reductions both sides could endorse. For all of this activity Goldin received mixed reviews: Some thought his approach was just the thing to make NASA soar again; others felt it bordered on the absurd. He promised to trim bureaucratic layers that created unnecessary delays. I thought faster, better, cheaper was a problematic slogan at the time because safety or reliability was no longer *primus inter pares*, the first among equals, in the three core values, or *topoi*, of the Old Technical Culture.

We now have a book-length treatment of the topic in Howard E. McCurdy's *Faster, Better, Cheaper: Low-Cost Innovation in the U.S.*

Space Program (2001). It came as a surprise to learn that the concept originated in the White House space policy staff during the Reagan administration. It was an attempt to hold down costs on the Strategic Defense Initiative. Daniel Goldin became an advocate of "faster, better, cheaper," as an executive at TRW. The concept came with Goldin when he became Administrator of NASA in 1992. Asked to cut costs at NASA, Goldin pushed faster, better, cheaper by launching smaller satellites in larger numbers. The concept was never intended to apply to manned flight, but it did have the effect of establishing different value premises for different projects.

His method was to encourage miniaturization of components, thus reducing the overall size of satellites. He saved money by reducing the size of project teams, or downsizing engineering personnel. At first the concept seemed to be promising. But 1999 was a bad year. Project managers using the approach had four failures in that year. McCurdy's assessment:

> In all, NASA officials attempted to fly 16 "faster, better, cheaper" missions between 1992 and January 1, 2000. The definition for inclusion in this group requires the project to have a substantially reduced cost ceiling, to be placed on a compressed development schedule, and to have someone with executive responsibility in NASA designate it as a "faster, better, cheaper" project. All of the spacecraft were small and most of them tested new technologies. Of the 16 projects undertaken, six had failed as of January 1, 2000. The resulting success rate, a paltry 63 percent, was significantly below spacecraft norms and well below the expectations for reliability that advocates placed on the initiative. (2001, 5)

We did, however, learn more about the relationships among the *topoi* of reliability, time, and cost. McCurdy summarizes studies that show that " 'faster' conspires with 'cheaper' to eliminate 'better' on complicated missions" (9). Efforts to reduce the risk of failure will increase costs and take more time. McCurdy explains what is called the "pick two" philosophy in this way: "According to this point of view, aerospace workers can improve two of the three terms in the 'faster, better, cheaper' equation, but not three. By the law of physics, according to this point of view, faster and cheaper cannot simultaneously be better" (9). McCurdy also quotes a researcher named David Bearden as saying that impaired performance, examined after the fact, are often found to be caused by " 'mismanagement or miscommunication,' difficulties traceable to budget ceilings or schedule constraints that are too tight for the project at hand" (11).

McCurdy argues that *complexity* is a crucial variable in the relationships among the *topoi* and to what he calls the "zone of reliability." For certain spacecraft

> the upper edge of the zone is bounded by cost and schedule goals ("faster, better"). The lower edge of the zone is created by complexity. Any attempt to reduce cost or schedule below the appropriate complexity level invites failure; any attempt to buffer failure with excessive spending (or time) violates the premise of the initiative. A narrow range exists between the two. (135)

The CAIB Board took a retrospective look at Goldin's tenure as Administrator (April 1, 1992, to November 17, 2001, the longest-serving person in the job) under this heading: "5.4 Turbulence in NASA Hits the Space Shuttle Program." The Board said Goldin engineered a torrent of policy changes, creating "not evolutionary change, but radical or discontinuous change," his tenure being "one of continuous turmoil, to which the Space Shuttle Program was not immune" (105). He was a follower of W. Edwards Deming, who had developed a set of management principles aimed at quality control while working in Japan. According to Goldin, one of those principles led him to argue that a corporate headquarters shouldn't try to exert "bureaucratic control" (105, a phrase we shall examine below) over a complex organization; instead it should set the strategy and delegate authority to operating units and give them the resources to get the job done.

He applied another of Deming's principles, the idea that organizational checks and balances are unnecessary—those doing the work should bear responsibility for quality. "It is arguable," wrote the Board, "whether these business principles can readily be applied to a government agency operating under civil service rules and in a politicized environment. Nevertheless, Goldin sought to implement them throughout his tenure" (106). He also sought to decrease the involvement of the agency's workforce in the Space Shuttle Program, making that engineering capability available to the International Space Station and the human exploration of Mars.

During his tenure Goldin reduced the workforce by 25 percent by means of attrition, early retirements, and cash bonuses for those who left NASA. There was also a hiring freeze most of the time, making it difficult to recruit new workers and young talent. Realizing he had gone too far, Goldin declared in June of 1999 that NASA faced a "Space Shuttle Crisis." By early 2000 NASA leaders became convinced that the downsizing had led to a skill imbalance and an "overtaxed workforce" (110). Problems of communication were identified at that time because of the complexity of the program and NASA's relationships with contractors.

Although the Report praises Goldin for some of what he tried to accomplish in his nine years at the helm, it is clear that the Board was uneasy with some of the business practices he introduced. In addition, his downsizing of the engineering staff dedicated to the space shuttle fleet made it difficult to close out, or categorize as solved, safety items. As for the cultural slogan, fasterbettercheaper, it contributed to the "production pressure" of keeping to an ambitious flight schedule.

The Safety Culture

The Board then sliced NASA's overall culture to examine the "safety culture," which they analyzed separately, and observed that it is "straining to hold together the vestiges of a once robust systems safety program" (177), reminding us again of Burke's mechanism of tragedy, a "straining" after purpose. A short history lesson makes clear that the safety culture had been weakened by government decisions over the years to reduce the workforce and make NASA more dependent on contractors for technical and safety support; this eroded NASA's in-house engineering depth, making it a slimmed-down agency largely run by contractors.

Section 7.3 compares NASA with "Best Safety Practices" of three other organizations, finding the former lacking. The first is the U.S. Navy's SUBSAFE, a program to prevent flooding in its submarines and to promote recovery of damaged subs by engineering depth, emphasizing individual as well as organizational responsibility and providing redundant and rapid methods of communicating problems up the line to decision makers. This program is similar to the redundant upward-directed channels of communication in the Marshall Center described in Chapter Four of this book.

The second organization using best safety practices is a joint Navy/Department of Energy organization responsible for all aspects of Navy nuclear propulsion; the nuclear Navy has 5,500 reactor years of experience—all of this without a single accident. They depend on several key elements: redundant paths of communicating problems; an insistence on hearing minority opinions; formal written reports; facing facts and details; and ability to manage change, including the obsolescence of warships.

The third organization with best safety practices they considered is the Aerospace Corporation, a federally funded Research and Development Center that supports the U.S. government in science and technology. It reports to the Air Force, giving independent verification that vehicles are ready for launch. The Air Force has, as a result, achieved a very low probability of failure in its launches using this independent verification system; the CAIB board thought this kind of system would provide NASA with mission assurance if tailored to the Shuttle Program.

Comparisons of these three safety systems with NASA's led the Board to declare NASA had a "broken safety culture" and a "flawed safety culture" on STS-107, the last flight of *Columbia*. Its vocabulary in supporting this diagnosis includes such expressions as "silence," "all voices must be heard," "stating preconceived conclusions," "play the devil's advocate and encourage an exhaustive debate," and many other terms focusing on communication and miscommunication. Its recommendations are written in the same vocabulary.

Chapter 8 of the CAIB Report

According to the article in the *New York Times,* either the Board or Admiral Gehman asked Diane Vaughan to write the eighth chapter of its report: "History as Cause: *Columbia* and *Challenger.*" Perhaps the reporter meant she wrote the first draft; the final report was written by professional writers. Chapter 8 does bear Vaughan's implicit signature. Eight of the 49 footnotes to the chapter contain references to Vaughan's work. One of the six headings, 8.2, also refers to her trademark phrase: "Failures of Foresight: Two Decision Histories and the *Normalization of Deviance*" (emphasis added).

8.1 *Echoes of* Challenger

This section quotes Sally Ride, the Board member who also served on the Rogers Commission, as saying there were "echoes" of *Challenger* in *Columbia*. The section details them, arguing that the history of O-ring incidents in the first accident were "normalized"; a similar process happened with foam debris incidents. Not everything in the organization was fixed or remedied after the first accident, and the stage was set for another accident. The penultimate paragraph of the section ends with these words:

> Third, the Board's focus on the context in which decision making occurred does not mean that individuals are not responsible and accountable. To the contrary, individuals always must assume responsibility for their actions. What it does mean is that NASA's problems cannot be solved simply by retirements, resignations, or transferring personnel. (195)

Recall that Chapter Five's review of two books about *Challenger* contrasted Jensen's stress on individuality and ethics with Vaughan's emphasis on structure, on the system: Blame the system, not the individual. Here the Board members are acknowledging that individual responsibility is important and, well, almost automatic. They want to add, however, that NASA's problems can't be solved by getting rid of certain people.

8.2 Failures of Foresight: Two Decision Histories and the Normalization of Deviance

This is a long and complex section showing that in both shuttle accidents the classifications of shuttle risks were downgraded over time from categories that mean dangerous to ones that are more acceptable. It concentrates on the meaning of the various terms used, even commenting on NASA's expression "in-family" as a "strange term indeed for a violation of system requirements" (196). Damage to the shuttle's Thermal Protection System was "normalized" by semantic moves even before launches of the craft began.

8.3 System Effects: The Impact of History and Politics on Risky Work

"The series of engineering decisions that normalized technical deviations shows one way that history became cause in both accidents" (197). This section spreads the blame, indicting the White House, Congress, and NASA leadership for making the agency politicized, vulnerable, and strapped for cash. The Shuttle Program "blocked the flow of critical information up the hierarchy" (197), a serious form of miscommunication. The checks and balances in the safety program weren't working. The dangerous parts of the NASA culture found in 1986 were present again in 2003, particularly the production pressure.

The chapter does use the word *communication*: "The Space Shuttle Program had altered its structure by outsourcing to contractors, which added to communication problems" (198). Again, in the same paragraph:

> When high-risk technology is the product and lives are at stake, safety, oversight, and communication flows are critical. The Board found that the Shuttle Program's normal chain of command and matrix system did not perform a check-and-balance function on either foam or O-rings. (198)

To save money NASA "attacked" its own safety system. The Board found shortages in safety personnel that were masked by a trick: Give one person multiple job titles. They found people who had more than one job title, including a case where one person was filling three positions, reporting to himself and making a foul compromise of what was supposed to provide checks and balances.

8.4 Organization, Culture, and Unintended Consequences

At the same time that NASA leaders were emphasizing safety, they were downsizing the safety personnel. Goldin's slogan of fasterbettercheaper combined with the cutbacks to send an ambiguous if not contradictory message to NASA personnel: *Efficiency is more important than safety.* In addition, the "can-do" culture of the Sixties was retained as a mere slogan, a no-longer-deserved cultural artifact. One symptom of this is NASA's stubborn insistence in calling the shuttle operational, like a school bus or delivery truck, when it wasn't. NASA also stopped listening to outside experts who warned them about their risks.

8.5 History as Cause: Two Accidents

This section places a series of five paragraphs from the CAIB Report in *italics*, interspersed with amplification and explanation in normal type, as a way of emphasizing conditions common to both accidents. They provide an excellent summary of the human factors as contributing causes to both accidents.

> *NASA had conflicting goals of cost, schedule, and safety. Safety lost out as the mandates of an "operational system" increased the schedule pres-*

sure. Scarce resources went to problems defined as more serious, rather than to foam strikes or O-ring erosion. (200)

This paragraph returns to the *topoi* discussed earlier; notice that safety or reliability is third. Schedule pressures interacted with scarce resources to produce a cultural crisis. This chapter does name a person in fixing a certain degree of blame. Linda Ham, who was the Mission Management Team Chair for the *Columbia* flight, is criticized for making "statements in many meetings reiterating her understanding that foam was a maintenance problem and a turnaround issue, not a safety-of-flight issue" (200). Ham is also mentioned as having a conflict of interest because she was scheduled for responsibilities for a subsequent flight, thus giving her a motive to avoid a slip in the schedule.

NASA's culture of bureaucratic accountability emphasized chain of command, procedure, following the rules, and going by the book. While rules and procedures were essential for coordination, they had an unintended but negative effect. Allegiance to hierarchy and procedure had replaced deference to NASA engineers' technical expertise. (200)

Again organizational communication—the chain of command—is employed to explain the *Columbia* accident. And the word *allegiance* in the last sentence is being used in the sense of commitment or identification because of the strong family resemblance noted earlier. Identification with hierarchy and procedure replaced deference to the engineering culture. Although the author of this paragraph unfortunately doesn't drive all the way to the ineluctable conclusion, evidence abounds in this chapter that it isn't a weakened or broken culture so much as it is *two antagonistic cultures.* There has been a reversal of the hierarchy, with the managerial culture establishing its dominance over the engineering culture, replacing the past deference.

In developing this point the author of this chapter of the CAIB Report mentions another kind of reversal common to both accidents: "Both *Challenger* and *Columbia* engineering teams were held to the usual quantitative standard of proof. But instead of having to prove that it was safe to fly, they were asked to prove that it was unsafe to fly" (201).

The organizational structure and hierarchy blocked effective communication of technical problems. Signals were overlooked, people were silenced, and useful information and dissenting views on technical issues did not surface at higher levels. What was communicated to parts of the organization was that O-ring erosion and foam debris were not problems. (201)

Here is additional evidence that there was a violent clash of cultures that produced blockages of communication, the silencing of members of the engineering culture; dissenting views were discouraged, not heard. Although the author does not go as far as I intend to go, it is clear the managerial culture *defeated* the engineering culture.

NASA structure changed as roles and responsibilities were transferred to contractors, which increased the dependence on the private sector for safety functions and risk assessment while simultaneously reducing the in-house capability to spot safety issues. (202)

In the *Challenger* case NASA was dependent on Thiokol for a risk assessment of the O-ring on the eve of the launch; in the *Columbia* case NASA was dependent on Boeing for the Crater analysis of foam debris damage. But there was a reciprocal dependence as well, in which the vendor tried to make the customer's management happy.

NASA's safety system lacked the resources, independence, personnel, and authority to successfully apply alternate perspectives to developing problems. Overlapping roles and responsibilities across multiple safety offices also undermined the possibility of a reliable system of checks and balances. (202)

This is somewhat repetitive but it does create a rationale for a recommendation the Board later makes about the independence of the safety program in NASA. Without being specific, this conclusion suggests that CAIB perceived groupthink and hindsight bias at work.

This is the last of the italicized sections of Chapter 8. There is a final section to examine.

8.6 Changing NASA's Organizational System

This section tries to summarize the chapter in a way that establishes criteria for organizational changes. It argues that NASA has little understanding of its own organization and problems. Production pressures must be reduced. And then this crucial and revealing sentence: "The same NASA whose engineers showed initiative and a solid working knowledge of how to get things done fast had a managerial culture with an allegiance to bureaucracy and cost-efficiency that squelched the engineers' efforts" (202–3).

The phrase "managerial culture" implies but again stops short of making explicit the notion of two antagonistic cultures. Allegiance again bears a family resemblance to identification with bureaucracy and cost-efficiency. Again there is evidence in support of a two-culture hypothesis, but the Board fails to take the next step. Instead, it moves on to criteria for organizational change, the first being that changes should be introduced with great care and toward the end of simplifying rather than complicating structure. The second criterion states that strategies for reorganization must be used to increase the clarity, strength, and presence of "signals that challenge assumptions about risk" (203). "Signal" is used here in this sense of argumentative communication. The final sentence of this section and the chapter is this: "Because ill-structured problems are less visible and therefore invite the normalization of deviance, they may be the riskiest of all" (203).

The Two Cultures Argument

With all due respect to the Board and its Report, and in the name of open communication, I offer a slightly dissenting interpretation of its analysis. After reading the whole report once and sections of it many, many times, I made another reading dedicated to testing—in a critical sense if not a scientific one—the thesis that its analysis of a broken or flawed culture can be made more specific, that the problem is one of *two antagonistic cultures, a managerial culture caught up in production decision premises, schedule, and budget, and dedicated to bureaucratic lines of command; the other, a subordinate engineering culture stretched too thin by attrition and intimidated by the management hierarchy.*

I now present the results of that second search in the form of quotations from the Report:

> The Mission Management Team declared the debris strike a "turnaround" issue and did not pursue [an engineering] request for imagery. (38)

> Within NASA centers, as Human Space Flight Program managers strove to maintain their view of the organization, they lost their ability to accept criticism, leading them to reject the recommenda-

tions of many boards and blue-ribbon panels, the Rogers Commission among them. (102)

Section 6.3 . . . reveals how engineers' concerns about risk and safety were competing with—and defeated by—management's belief that foam could not hurt the Orbiter. (121)

What drove managers to reject the recommendation that the foam loss be deemed an In-Flight anomaly? . . . As will be discussed in Section 6.2, the need to adhere to the Node 2 launch schedule also appears to have influenced their decision. (125)

With no engineering analysis, shuttle managers used past success as a justification for future flights, and made no change to the External Tank configurations planned for future flights. (126)

It is significant that in retrospect, several NASA managers identified their acceptance of this flight rationale as a serious error. (126)

Shuttle Program managers appear to have confused the notion of foam posing an "accepted risk" with foam not being a "safety of flight issue." At times, the pressure to meet the flight schedule appeared to cut short engineering efforts to resolve the foam-shedding problem. (130)

The Board posed the question: Was there undue pressure to nail the Node 2 launch date to the February 19, 2004, signpost? The management and workforce of the Shuttle and Space Station Programs answered the question differently. Various members of NASA upper management gave a definite "no." In contrast, the workforce within both programs thought there was considerable management focus on Node 2 and resulting pressure to hold firm to that launch date, and individuals were becoming concerned that safety might be compromised. *The weight of the evidence supports the workforce view.* (131, emphasis added)

Flight Readiness Review charts [for *Columbia*'s last flight] provided a flawed flight rationale by concluding that the foam loss was "not a safety flight" issue. Interestingly, during *Columbia*'s mission, the Chair of the Mission Management Team, Linda Ham, would characterize that reasoning as "lousy"—though neither she nor the Shuttle Program Manager, who were both present at the meeting, mentioned it at the time. (137–138)

I pause here to consider the meaning of this sample of nine quotations from the CAIB Report restated in condensed form: Managers failed to pursue engineers' concerns; managers were unable to accept criticism, even by the Rogers Commission; engineers' concerns were "defeated" by managers' beliefs; managers rejected engineers' requests because of the schedule; managers relied on past successes, not engineering analysis; several managers admitted to a serious error in their flight rationale; managers seemed confused about the meaning of their categories of flight-risk assessment; managers and engineers disagreed about the pressure on the latter created by the Node 2 flight scheduled in 2004, with the Board deciding the engineers were correct; at least two managers thought the flight rationale was "lousy" at the time it was presented and yet they used that rationale to squelch inquiries and requests for more imagery.

I submit that if these claims support the idea of a broken culture, it was a culture broken into two pieces, two antagonistic cultures. Many, many more quotations could be adduced to prove beyond doubt from the Report's own observations that the two-culture hypothesis is valid. Instead, I shall present only one more quotation and rest my case: "The January 27 to January 31, phone and e-mail exchanges, primarily between NASA engineers at Langley and Johnson, illustrate another symptom of the 'cultural fence' that impairs open communication between mission managers and working engineers" (169). A "cultural fence" between two groups is a clear sign of the existence of two different cultures.

Let's see what happens to the notions of a broken culture and two antagonistic cultures by running through the matrix of cultural dimensions of the Old Technical Culture as established by McCurdy and modified by me.

1. Research and Testing

On this aspect we can definitively say that the OTC has weakened or been broken. Neither the O-ring nor the foam shedding was tested adequately; tests had to be ordered by the Board after the accident. But it is the engineering culture that designs and executes the tests. There are many fewer NASA engineers now than in the glory days, and cuts were made in part by management decisions to reduce the workforce and rely on contractors. The management culture weakened the technical culture.

2. In-house Technical Capability

Again it is obvious that the OTC has been weakened on this dimension of culture. The weakness accrued, however, from the re-

duction in the engineering culture as a result of decisions made by
the management culture.

3. Hands-on Experience

This dimension has diminished and again falls more heavily on
the engineering community and culture than on the management
culture. Even in the OTC there was a tension between management
and the engineers on this matter because engineers kept their hands
dirty while management kept their hands clean shuffling paper.
There are far fewer personnel now in the engineering ranks, reducing automatically and by definition the number of dirty hands.

4. Exceptional People

Fears were expressed in its second two decades that NASA had-
n't been able to replace the able young people of the Sixties who retired. The relatively routine nature of shuttle flights is certainly less
exciting than the visionary days of the first decade in NASA. There
are fewer engineers today in NASA, but I know of no objective mea-
surement of their ability. The CAIB Report does refer to one experienced observer of the technical workforce, Beth Dickey, who referred to them as "The Few, The Tired," in 2001, suggesting that
downsizing and budget cuts left NASA with the job of exploring the
universe with a smaller, less experienced staff (118). The criticisms
of the CAIB Board are more sharply pointed at management than at
the engineering culture, making one doubt there is the level of management competence known in the glory days.

5. Risk and Failure

The risk is still high, even though NASA management declared
the shuttle operational before the *Challenger* accident. Two catastrophic accidents that killed 14 astronauts would suggest that the
failure rate is higher than before. The two official reports have been
more critical of the management culture of NASA than of the engineering culture, although both get some blame. In the case of the
Challenger explosion it was the engineering culture that feared the
risk of flying without test data; NASA managers wanted to fly without the data. In the case of *Columbia* managers prevented engineers
from obtaining the images they needed to do a sound engineering
analysis. There is evidence here for two cultures and tension be-

tween them, the managers being more driven by the schedule than the engineers.

6. Open Communication

Under this dimension we see the severest criticism of NASA in both official post mortems. In the second case the Board directed a withering criticism of mission managers. They did on a couple of occasions decry the "reticence" of engineers in speaking up for what they believe in, but management's record is deplorable. Here we find the best evidence for the "fence" between the two cultures.

Fault found with communication in NASA by CAIB is almost completely placed with management, a stance or position that could hardly have been avoided. It is the function of management to create and maintain the channels of communication, particularly vertical channels, in both directions. One of the clearest statements of this principle was articulated in the classic text on organization and management, *The Functions of the Executive,* by Chester Barnard (1938): "The elements of an organization are . . . (1) communication; (2) willingness to serve; and (3) common purpose" (82). The first executive or management function is, therefore, to create and maintain the system of communication. As we have seen from a variety of sources and evidence, vertical communication—both up and down—is absolutely crucial in high-risk organizations; in fact we can say with certainty that in the best practices, including the Marshall Center during the first decade of NASA, *redundant channels of vertical communication were a basic necessity and were effectively used.*

The language/meaning dimension of communication and culture was stressed in Chapter Three. The managerial/bureaucratic culture was expressed in the language or dialect of schedule and cost and in categories of risk that managers confused; it enforced the rules and chain of command, even though managers skipped most of their required meetings. The engineering/concertive culture is still expressed in quantification and test results. These two different dialects helped create the cultural fence between the two groups.

Finally, the Board says "that NASA is not functioning as a learning organization" (127), a reminder of the interpretation Weick and Ashford (2001) made of my description of the Monday Notes—that this method of frequent communication about engineering problems up and down the organization made MSFC in its first decade a paradigm case of the learning organization.

7. Organizational Identification and Control

As noted previously, I began to study organizational identification when I first went to work for NASA at the Marshall Center in 1967. Never since then have I either experienced or observed such high levels of organizational identification. We were all in it together, everyone working for the good of the Apollo Program. The open vertical channels allowed young engineers to take on the whole hierarchy of the center; this communicative behavior was not only tolerated, it was encouraged and appreciated—and sometimes that lonely engineer or small working group turned out to be right. The long history of attrition and absence of an inspiring vision since then in NASA have reduced this level of commitment.

The Board again has provided us with relevant evidence on this dimension:

> NASA's culture of bureaucratic accountability emphasized chain of command, procedure, following the rules, and going by the book. While rules and procedures were essential for coordination, they had an unintended but negative effect. Allegiance to hierarchy and procedure had replaced deference to NASA engineers' technical expertise. (200, emphasis in original)

Allegiance is a word that bears a family resemblance to identification, or as perhaps it could be expressed in NASA's jargon, the two concepts are "in family." The managers now identify with their long-term schedule, their hierarchy, rules, and chain of command rather than with in-house technical competence. That is the *new management culture*. The engineers, on the other hand, began to de-identify—became increasingly alienated is a better way to express it—from the organization as manifested by the rules and bureaucratization, concerns of cost and schedule, and an increasing lack of flexibility and agility. In short, they were alienated by and from the management culture, making communication up and down the line more problematic.

A Footnote on Control

This process of alienation, the dialectical opposite of identification, has tremendous implications for the problem of control. The theory of organizational control proposed by Tompkins and Cheney (1985) took as its historical foundation the work by an economist, Richard Edwards (1978). Edwards claims that the control of organizations has changed over the years. It has in all forms needed three elements in a closed loop of communication: direction, monitoring and feedback, and redirection. Historically, in its first manifestation there was *simple* control, the supervision of the work by an owner, and later by his overseers and foremen. As organizations grew in size, foremen were removed from the owners sufficiently to abuse the workers, often physically, and by asking for bribes in return for jobs. Workers were naturally alienated, rebelled against the abuse, and Congressional hearings were held to investigate the abuse of simple control.

Owners and professional managers have always looked to reduce workers' resistance to control; they seek and create less obtrusive methods. The next step was to create *technical* control, the use of machinery to replace the overseers and foremen. Edwards traced the origin of technical control back to meat-packing plants in England; the carcasses of cattle were hoisted onto hooks on a moving cable. Stationary workers chopped parts off the animal as it moved along. Hence the first assembly line is more accurately called a *disassembly* line. It took an American genius to prove the system applicable to many forms of manufacturing.

When Henry Ford began to build automobiles on an endless conveyor—stationary workers adding parts rather than subtracting them—we can truly call it an assembly line. Productivity and efficiency soared, and Ford was able to *reduce* the ratio of foremen to workers to an amazing 1:58, amazing when the classic span of control held the optimal ratio to be 1:7. Control was maintained in part by the assembly line or conveyor belt because it established directions by dictating the pace of the work. The line's rate of speed brought the "orders" to the worker. Fewer foremen were needed to give orders, monitor, and discipline the workers. It seemed a miraculous achievement in terms of efficiency, but over time workers found reason to resist technical control.

· As technical control spread, owners and managers sought to increase productivity by speeding up the line. The faster the line, the harder the work, of course, and the more dangerous the working conditions became. Workers began to feel more like cogs in a machine than human beings crafting a product. They were also deskilled; iron workers knew the secrets of their work, owners and managers didn't. Owners and their managers knew the secrets of making steel, and so the workers felt like unskilled robots. They did learn that they could sabotage production on the line and stop production by the sit-down strike, in which they sat down at their stations and refused to work. The United Auto Workers Union was organized, forcing owners to enter into collective bargaining about wages and working conditions.

Meanwhile, improved communication such as telegraph and telephone allowed organizations to grow because the control loop could transcend long distances. As organizations grew, particularly in nonmanufacturing areas of the economy, a new form of control came into existence: _bureaucratic_ control. The essence of the new system, foreseen by the German sociologist Max Weber (1870–1922), was described in its pure type as characterized by rationality and rules, by trained and salaried experts who applied the exhaustive rules to specific cases that appealed to them. We tend to forget that bureaucracy was an improvement over the organizations that preceded it. The Civil Service System in the U.S. government was much better than the patronage system before it, in which friends and relatives could be appointed to positions in which they made decisions not by the rules but by doing favors for friends. Corruption and cronyism abounded; bureaucracy reduced personal whim by its cool rationality and a do-it-by-the-numbers approach.

I probably don't have to mention that workers in bureaucracies, and their clients, began to chafe under this system of control. They objected to the rigidity of the rules, the goal displacement in which the rules became more important than the purposes and goals of the organizations. A new phrase appeared in our language, the "red tape" of rules and regulations; it became an iron cage. Bureaucracies and their bureaucrats realized that they needed more flexibility— less centralization and more decentralization.

Although businesses adopted bureaucratic control, we typically connect bureaucracy with government agencies and their "bureaus," or subdivisions. The Post Office and the military come to mind because of their rules and regulations. That is why I, as a student of organization and communication, was startled when I walked into the Marshall Space Flight Center in 1967. Although there was much paperwork to be done to hire me as a Summer Faculty Consultant in Organizational Communication—I was classified as GS-13—part of that was required for my security clearance. I saw activities that amazed me. I saw much more decentralization than expected. I heard of Working Groups and self-created Teams that wrote their own rules, or to mix metaphors, picked up the ball and ran with it.

Take the Monday Notes, for example, as a counter to the impersonality of bureaucracy. There was no NASA-MSFC form with letters and numbers, simply one page of paper with a person's name and date on it. The note vaulted over layers in the chain of command; it came back with notations by von Braun on the notes of all the participants. Arguments flew horizontally between labs and program offices and in both directions vertically, along with praise and promptings. I was allowed to interview anyone I wanted—to enjoy what seemed to be complete access to the organization.

The pride, loyalty, allegiance, and commitment—the whole family of terms I chose to call organizational identification—was observable. People talked about it. Oh, there were complaints about paper shufflers in the personnel office, what we call human resources today, because of their rules and regulations, but there was clearly a feeling of all for one and one for all. And as I mentioned in Chapter Four, it came over me as well. I felt a sense of pride and identification with NASA and MSFC and the Apollo Program; it became part of my identity and I was part of it.

For the rest of my career I would work on the theory of organizational identification. In the early 1980s I saw its connection with control, but it was a different kind of control than the three stages of control outlined by Edwards. It was more democratic. Groups solved problems in a face-to-face setting, applying factual and value premises to problems. We could see some trends in private corporations that fit with what I had seen at MSFC: teamwork. It was decentralized in that teams had authority to make decisions and establish their own procedures in group discussions. George Cheney worked on the theory with me.

We needed a name for the new form of control. We turned to a close colleague, Elaine Tompkins, for a name. She coined a word: *concertive*. (My brand-new computer with up-to-date software still underlines it in red because it is a neologism unknown to it.) She meant that groups act flexibly in *concert* to solve problems and get the work done. We predicted that it would ironically increase the total amount of control in the system because every participant was an informal supervisor of the group, correcting each other and the group itself if necessary. James Barker, now in the Management Department of the Air Force Academy, conducted award-winning empirical work validating that claim of the theory (1993).

I return now to my analysis of the CAIB Report's analysis of NASA's organizational problems: The Board found that the Safety Program was "unnecessarily bureaucratic and defeats NASA's stated objective of providing an independent safety function" (186). And as we saw earlier, the Report concludes that "NASA's culture of bureaucratic accountability emphasized chain of command, procedure, following the rules, and going by the book. While rules and procedures were essential for coordination, they had an unintended but negative effect" (200). They go on to say that the bureaucratic hierarchy blocked effective communication; dissenting views did not surface.

I conclude that NASA had lost some of its original decentralized, concertive control and moved backward to a greater reliance on bureaucratic control. This is implicit in the recommendations of the Board, as will soon be seen. All of this is consistent with the notion of two cultures with a fence between them. Call the one a managerial/bureaucratic control culture, the other a technical/concertive control culture. Recall that the engineering group of Thiokol stood together in recommending "no go" on the eve of the *Challenger* launch; recall that the engineering groups in the Intercenter Photo Working Group and the Debris Assessment Team requested additional imaging. These concertive groups were overruled by the managerial/bureaucratic control culture. My final piece of evidence is this hard-hitting judgment of the Board on this matter:

> The same NASA whose engineers showed initiative and a solid working knowledge of how to get things done fast had a managerial culture with an allegiance to bureaucracy and cost-efficiency that squelched the engineers' efforts. When it came to managers' own actions, however, a different set of rules prevailed. The Board found that Mission Management was not able to recognize that in unprecedented conditions, when lives were on the line, flexibility and democratic process should take priority over bureaucratic response. (203)

The words to reflect on in that extraordinary statement are "managerial culture" and "allegiance to bureaucracy" and "cost efficiency"; they stand in opposition to the engineering/concertive culture characterized by "initiative," "how to get things done fast," "flexibility," and "democratic process." Unfortunately the one cultural cluster defeated the other.

Recommendations

It is somewhat difficult to summarize the Board's recommendations because they are presented in three different ways: the first is in the unfolding chapters of the report, along with findings and observations; later the recommendations are organized in two different ways. The first method is in Chapter 9, "Implications for the Future of Human Space Flight," in which the Board looks to the future in three timeframes: (1) near-term, NASA's return to flight; (2) mid-term, finding a replacement for the shuttle; (3) long-term, future directions for the space program. The second method is by subject area, such as Thermal Protection System, Bolt Catchers, Organization, and many others. I shall introduce the timeframes briefly and concentrate on the topical recommendations.

1. Near-Term: Return to Flight

The Board supports returning the shuttle to flight as soon as possible with an overriding consideration for safety. They list five technical areas for intensive work, including such options for survival as crew escape systems and safe havens. But the essence is organizational in nature, recommending an independent Technical Engineering Authority and an independent safety program.

2. Mid-Term: Continuing to Fly

The Board again stressed organizational changes in several parts: "One is separating technical authority from the functions of managing schedules and cost. Another is an independent Safety and Mission Assurance organization. The third is the capability for effective systems integration. [This is another echo from the past: The goal of the reorganization of MSFC I worked on in 1968 was to increase the capability for systems engineering and systems integration.] Perhaps even more challenging than these organizational changes are the cultural changes required" (208), topics considered below. The Board also called for the recertification of the shuttle, a process of making sure the craft has not exceeded its original design life because of its age.

3. Long-Term: Future Direction for the United States in Space

Before the Board could take a long look forward, it had to take a long look backward in time, finding that over the prior three decades NASA had not received a national mandate since President Kennedy's charge to send Americans to the moon and bring them back. The agency had to scramble for its very existence, diversifying its projects and missions: "The result, as noted throughout Part Two of the report, is an organization straining to do too much with too little" (209), straining after purpose being the mechanism of tragedy.

Throughout the report the Board is critical of both Congress and the White House over the past three decades for a lack of a national vision for space and the lack of sustained government commitment, particularly over the past decade. "It is the view of the Board that the previous attempts to *develop a replacement vehicle for the aging shuttle represent a failure of national leadership*" (211, emphasis in original). The failure was due to expectations for major technological advances that didn't materialize. The Board therefore recommended the development of a New Space Transportation System, the replacement of the shuttle.

Chapter 11 is called "Recommendations" and is organized by three headings: Part One—The Accident; Part Two—Why the Accident Occurred; Part Three—A Look Ahead. The lead paragraph of the chapter is as follows: "It is the Board's opinion that good leadership can direct a culture to adapt to new realities. NASA's culture must change, and the Board intends the following recommendations to be steps toward effecting this change" (225).

Curiously, the words *leadership* and *culture* do not appear in the recommendations, nor does *communication*. Perhaps there is an implication for insiders that heads should roll—that leaders should leave and new ones be brought into prominence. In fact there were some resignations and transfers after the accident, consistent with the precedent set after *Challenger*. The only recommendation that smacks of communication and culture is the one dealing with training. The Board recommended that the Mission Management Team should have expanded training in which they face potential contingencies, i.e., safety risks, to the crew and vehicle, and in which they are required "to assemble and interact with support organizations across NASA/Contractor lines and in various locations" (226). Interact as a word has a family resemblance to communicate.

Recommendations are also organized by subject matter, from hardware, such as bolt catchers and RCC, to training and, most important to this analysis, Organization. The Report repeats the recommendation noted above, namely, to establish and define the responsibilities of an Independent Technical Engineering Authority funded directly from NASA Headquarters, an independent safety program, and a reorganized Space Shuttle Integration Office to make it capable of, well, integrating all aspects of the Space Shuttle Program. NASA is also told to submit annual reports of its implementation of the recommendations to Congress as part of the budget review process.

It is curious that although the Board finds fault with the Problem Reporting and Corrective Action System and reaches findings about communication and culture, they make no specific recommendations about these topics. They must have believed (1) that their diagnosis of these problems would be enough to see them solved, or (2) that they were inevitable and insoluble, or (3) that their recommendations of reorganization would eliminate or reduce them. Then it *dawned on me*. The recommendations are hard to understand because the Board *didn't use fault-tree analysis* on the issues of organization, culture, and communication. At least it is explicit neither graphically nor discursively. Perhaps they tried collectively to emulate Sherlock Holmes in this section of the Report, by deducing the answers in their head. It is also difficult to construct a fault-tree of my own because I don't know what branches they considered, if any, before eliminating them. They do say they concluded that the accident was "probably not an anomalous, random event" (9). Perhaps they rejected arguments opposite to what they included—null hypotheses—such as (1) communication was not a problem, (2) the OTC was still strong, and so on.

I'll try to infer their reasoning process by examining the rationale for the relevant recommendations. Much of the rationale appears on pages 208 and 209 of Chapter 9. The arguments surrounding the organizational changes are developed tersely in several ways. Despite an economy of discourse, they do tell us something about the Board's reasoning. They want NASA to adopt the characteristics of high-reliability organizations, the best practices discussed above. What are the characteristics? "One is separating technical authority from the functions of managing schedules and cost." It is clear by this sentence that the Board wants to strip shuttle management of the ability to make technical decisions. The change will restore authority to the engineering/concertive culture, to the working groups and ad hoc peer reviews and research and development teams. "Another is an independent Safety and Mission Assurance organization." Independent of what, we might ask, and the answer can only be independence from the managerial/bureaucratic culture. The safety system was passive and without authority during the *Columbia* travail. "The third is the capability for effective systems integration." Systems integration is the function of taking into account all the components of the shuttle, how they affect each other and the spacecraft as a system. The managerial/bureaucratic culture was too narrow, too oriented to this or that possible component, not how it could indirectly affect flight safety.

"Perhaps even more challenging than these organizational changes," continues the Board, "are the cultural changes required. ... The Board's view is that cultural problems are unlikely to be corrected without top level leadership." New leaders will have to be found to solve the cultural problems. The new leaders "will have to rid the system of practices and patterns that have been validated simply because they have been around so long" (208). The Board then gives examples of these practices and patterns:

> the tendency to keep knowledge of problems contained within a Center or program; making technical decisions without in-depth, peer-reviewed technical analysis; and an unofficial hierarchy or caste system created by placing excessive power in one office. (208)

These practices are mainly found to be in the managerial/bureaucratic culture, not in the engineering/concertive culture.

"Such factors interfere with open communication, impede the sharing of lessons learned" and others. "Collectively, these undesirable characteristics threaten safety" (208). The Board thus does care about culture and communication, but their informal, nongraphic fault-tree analysis puts these factors under the branch of leadership, a new leadership that can root out the patterns and practices of the managerial/bureaucratic system.

The final inference is that the stress on independence for a technical authority and for safety and flight assurance and independence from the time and cost *topoi* of the managerial/bureaucratic culture are designed to set up the checks and balances of the Old Technical System. This would implicitly mean a reversal of the attempt by Administrator Goldin to eliminate such checks and balances. This should turn out to be another way in which to increase reliability and safety. Collectively, the organizational changes are intended to *reverse the destructive hierarchical relationship between the managerial/bureaucratic culture and the technical/concertive culture.* Thus, the organizational changes help the change in leadership to reverse the dangerous hierarchy NASA had allowed to emerge during the reduction of the engineering workforce.

The Board is not overly optimistic about their recommendations. They admit they will be difficult to initiate and will "encounter institutional resistance" (209). Nevertheless, the Board wants them carried out completely. Ultimately, NASA is an independent agency answering only to Congress and the White House. Having been critical of how those two institutions dealt with NASA in the past, the Board seems to put them on notice, to alert them to the need to monitor NASA's compliance during its annual budget reviews.

My own view is that the Board's recommendations might have been more graphic and explicit. I sense they underestimate several obstacles to the adoption of the recommendations and to the achievement of a new and effective NASA. One is that NASA is unlikely to improve the quality of the engineering/concertive culture without a massive infusion of new personnel, to bring the technical strength back far enough to achieve the near-, mid- and long-term goals. That would take a lot of money and a lot of recruitment. I think it's possible that they may have underestimated the possibility of improving the leadership enough to bring about the *necessary changes in culture and communication.* They may well underestimate the amount of resistance the managerial/bureaucratic culture will summon to resist the changes.

I do applaud their efforts to make systemic changes. In addition, I wish to highlight comments made by the Board that are consistent with my reservations in the previous paragraph. By drawing parallels between the two shuttle accidents and by stressing systems and structures, the Board

> does not mean that individuals are not responsible and accountable. To the contrary, individuals always must assume responsibility for their actions. What it does mean is that NASA's problems cannot be solved simply by retirements, resignations, or transferring personnel. (195)

This statement is an echo of what Claus Jensen said about the meaning of the *Challenger* in his book *No Downlink*. It is an echo as well of Automatic Responsibility, that principle of the Old Technical Culture at the Marshall Center that placed responsibility with every individual for either fixing problems or reporting them upward. Perhaps it should be reintroduced in the attempt to create a new culture in NASA. The CAIB does use communication as a central term in its diagnosis, does condemn the reversal of the burden of proof in both accidents, and does remind us that individual responsibility is important to the success of high-risk enterprises.

Finally, it's doubtful whether the Board had much of a choice about returning the shuttle to flight—as soon as safety issues permit—because of the shuttle's co-dependent relationship with the International Space Station. Predictably, they do want NASA to work to anticipate what the next O-ring or foam debris problem will be. They hope to achieve greater safety by reorganizing what I have called the two antagonistic cultures, by reversing the dangerous hierarchy, by giving greater authority to the engineering culture. In addition, by creating an independent safety program the Board is saying the current managerial-bureaucratic method of control has failed. The independence of the refigured engineering and safety systems means the reintroduction of a system of checks and balances that Administrator Goldin sought to remove.

The mysteries have been solved. This is not, however, the end of the story. Prior to and after the *Columbia* accident, a failure on NASA's part, other organizations—religious, business and government—were also suffering serious failings. In some cases they collapsed. The following chapter searches through those organizations for evidence that there were some common causes in the decline of so many U.S. organizations and institutions. ✦

Chapter Eight

The *Challenger* Syndrome and the Decline of American Organizations and Institutions: 'Speaking Truth to Power'

> I say we had best look at our times and lands searchingly in the face, like a physician diagnosing some deep disease. Never was there, perhaps, more hollowness at heart than at present, and here in the United States. . . . The spectacle is appalling. We live in an atmosphere of hypocrisy throughout. . . . A lot of churches, sects, etc., the most dismal phantasms I know, usurp the name of religion. . . . The depravity of the business classes of our country is not less than has been supposed, but infinitely greater. . . . In business (this all-devouring modern word, business), the one sole object is, by any means, pecuniary gain.
>
> —*Source revealed within this chapter*

Nearly everything the great Irish writer James Joyce wrote, including such classics as *Dubliners, Portrait of the Artist as a Young Man, Ulysses,* and *Finnegan's Wake,* was set in Dublin, his hometown. This is true even though Joyce left Dublin early in adulthood to devote his life to art in Europe—for reasons made understandable in the autobiographical *Portrait*—in such cities as Paris and Trieste. How can we account for this fixation on one city? He even referred to his hometown as "Dear Dirty Dublin," set in a country, Ireland, he described as the old sow that devours her own farrow, or piglets.

Joyce believed that if he could get to the heart of Dublin he could get to the heart of all cities. He used to claim that if a catastrophe demolished Dublin, it could be reconstructed—circa 1900—by using his novel *Ulysses* as a blueprint. The banks are there, Trinity College, the newspaper office, the pubs, and the churches. But getting to the heart of Dublin also meant portraying the culture of the city, the language of the various institutions such as education, art, religion, and commerce, as well as the defining activities of rhetoric, sex, and parenting.

The first seven chapters of this book differ from those in most textbooks by concentrating on one organization over an unusually long period of time instead of trying to express abstractions about *all* organizations. Could it be that by looking closely at one organization over time we might learn about all organizations? I can hear the reader's silent answer: not likely. But we have learned quite a bit about organizations by concentrating on NASA. Let me illustrate. We learned that organizations can be established in two ways, either by combining existing organizations into a new entity or by creating wholly new organizations. NASA used both methods. Either method involves the creation of a new system of communication by which the organization functions. The openness and candor in the culture allowed the informal system of communication to supplement and complement the formal system of official channels and organization charts. We saw also how organizations adapt to the environment by mitosis, the division of one organizational unit into two.

We saw an organization in a life cycle, from birth through its period of rapid growth to its precocious period of brilliance—Apollo. During these glory days MSFC achieved something close to the Ideal Managerial Climate. We saw its subsequent period of shrinking and shriveling as it lost its strength, sense of purpose, and value premises. Unhappily we saw it experience two mind-numbing, humiliating disasters. That allowed us to watch culture and communication and leadership variables correlate with the ups and downs of the organization. The individual was silenced by bureaucratic management. We also saw a manifestation of an organization's tendency to protect and defend itself from outsiders.

We learned how to define and "read" a culture, how it is manifested in communication and meaning, how it can be strong, influencing actions and decisions to achieve excellence. We also saw how it can weaken and how communication practices can change. We can now, looking back in time, see how a second accident was heard as an echo. That is, *Columbia* was an echo of *Challenger,* proving how powerful *institutional inertia* involving bad habits of communication can be. Organizational *praxis,* the everyday practice of communication, can become so habitual that it is difficult to change. So it is with culture.

We saw that a culture can divide into two antagonistic cultures, two warring tribes with a cultural fence between them: in this case the managerial/bureaucratic subculture and the weakened engineering/concertive one. We watched a reversal of status, as the engineers became second-class members, forced to communicate through the formal channels, intimidated by the managers. The informal system of communication could no longer save the formal system. Engineers found themselves required to prove spacecraft would *not* fly, a reversal of the normal, healthy, safe tradition of the burden of proof, of being expected to prove it will be successful.

We also learned how the ultimate decision premises—the *topoi*—varied with the changes in culture and communication. During the glory years of the organization it succeeded by relegating schedule to a priority below safety. There were many decisions to "scrub" or delay a flight. Dr. von Braun once told me some of these decisions were found to be based on such "minor" matters as a faulty fan belt in a tracking station in, say, Bermuda. As the culture weakened and communication practices were reversed, the schedule—*production*—became the most important premise or *topos.* Communication practices became "unethical" in that they were contrary to the earlier commandments of safety and respect for human life. Managers silenced engineers who raised doubts. The "oughtness" of the old technical culture was violated. Engineers were afraid and unable to stand up and express concerns; reviews became rituals instead of open communication. Indeed, open communication—particularly upward communication—was transformed into what Stanley Deetz (1992) calls *discursive closure*—the opposite of openness and candor.

We have seen the tension between the individual and the system, between the worker and the structure, the employee and the hierarchy, between human beings and the cultural webs they spin like spiders. Cultures affect individuals and individuals influence cultures. In the end, however, it is the individual who must be upright, who must stand up against the system when it violates the ultimate values. The system cannot save itself. This is the *Challenger* Syndrome. Perhaps now we should call it the *Challenger-Columbia* Syndrome because of the echoes and parallels. We also learned a highly important lesson—that disciplined communication practices, such as the Monday Notes, made MSFC a *model of the learning organization.*

In the section that follows I shall introduce some contemporaneous organizations and institutions that are in trouble; some are in worse shape than NASA and a few have suffered the ultimate demise—they no longer exist! This should be no surprise, for as Barnard (1938) observed, most organizations die. Few organizations last more than a year or two. There are precious few organizations that have lasted for 200 years. As we move through other, shorter case studies, the purpose is to look for similarities; how do they differ and how are they like NASA? We will be testing our hypotheses, not in a strict scientific sense as explained earlier, but with the highest rigor known to the humanities and interpretive social science. We should be on the lookout for evidence that the organization "cut corners," put premises or *topoi* such as *production* or *profits* or *cost-cutting* above the ultimate values of safety and reliability and honesty. We should also be equally concerned about communication—the ethics of communication. We have seen that open communication is a correlate of organizational success in the case of NASA. Discursive closure and deceptive communication can lead to disaster.

I do not want to give the impression that the organizations I've singled out for criticism represent the lowest point in the history of organizations; I will not be discussing slavery or child labor or the burning at the stake of those whose religious views were different from what was orthodox at that moment. Nonetheless, the United States claims to be more enlightened, more compassionate today than in the past. We therefore ought to be critical of organizational practices that fail to live up to enlightened standards. In addition, criticism is one of the ways in which to get both individuals and organizations to improve their actions.

I

On Thursday, January 9, 2003, just seven days before the last launch of the shuttle *Columbia*, the Public Broadcasting System performed a great public service by presenting a *Frontline* show called "A Dangerous Business." I turned on the television set in time to see much of it and knew immediately it was a story that was relevant to the thesis of a keynote address I was scheduled to give nine days later on January 18, 2003. The next day I got a transcript of the program and learned that it was the culmination of nine months of joint investigations by PBS, the *New York Times*, and the Canadian Broadcasting Company. An article published in the *New York Times* the next day was also part of the investigation into one of the most dangerous companies in America: the McWane Corporation, founded in Birmingham, Alabama, in 1921, by J. R. McWane; the current head is Phillip McWane, great-grandson of the founder. The McWanes are "one of the wealthiest families in the south, known for their philanthropy, endowing museums and universities. And they are also known for their avoidance of publicity" (*Frontline*, 6). The company makes sewer pipes in foundries scattered over ten states and Canada, "and over the past seven years the company has amassed more safety violations than all its major competitors combined" (*Frontline*, 1).

Until the journalistic investigation no outsiders knew of all the company's violations, which had been collected in various government offices around the country—there is no central file of these violations. McWane has received notice of more than 450 environmental violations and 400 safety violations since 1995: respirators aren't provided to workers; machines lack safety guards; employees are inexperienced and untrained; flammable liquids are mishandled; forklifts don't have brakes. Nine workers have been killed in McWane's plants since 1995; at least three of these accidents were caused by deliberate violations of Federal safety standards, and safety lapses account for at least five other deaths. Countless other employees have suffered burns and amputations, and one worker, monitored closely because of management suspicions that he was a malingerer, was finally diagnosed after x-rays of having a broken back. Investigations of these accidents rarely go anywhere. Local district attorneys typically refer the cases to the Occupational Safety and Health Administration; OSHA cannot prosecute—it can only refer cases it believes to be willful violations to the Justice Department. Only one of the nine deaths was referred to Justice, and the outcome was a single misdemeanor plea. No executives were charged. In addition to the injuries and deaths, McWane is a major polluter. Foundries use a tremendous amount of water; their wastewater is toxic and it is supposed to be put in holding pools. McWane knowingly used pools that were too small, causing overflow that washed the toxins into streams, storm sewers, and the Delaware River. McWane has also been sued for polluting emissions.

Let's consider more carefully one of the deaths: Frank J. Wagner, 40, was an employee at McWane's plant in Elmira, New York. On January 13, 1995, he was ordered to incinerate hundreds of gallons of old industrial paint, highly volatile stuff; he was told to burn it in an oven not built for such a purpose. It became a bomb, of course, and killed Mr. Wagner. (This case recalls the fire investigations made by Benjamin Lee Whorf but is likely to yield only ethical problems, not linguistic principles.) By late 1995 prosecutors felt they had a slam dunk case against McWane and prepared draft indictments accusing the company and several executives "with a string of felonies, including second-degree manslaughter, endangering public health and unlawful hazardous waste" (*New York Times,* 9). They asked the new attorney general of New York, Dennis C. Vacco, for a Grand Jury. But Vacco delayed seeking an indictment so long that the prosecutors gave up and turned the evidence over to Federal prosecutors. The investigation later turned up a letter McWane had sent Vacco, containing a threat to the attorney general that his political career would be destroyed if he sought indictments. It said that such an action would show him not to be "pro-business," and that McWane's closing of the Elmira plant would mean the loss of 320 jobs in an economically depressed area.

Let's now concentrate for a few moments on how McWane operates. They buy antiquated plants and increase profitability through what they call "Disciplined Management Practices"; at the Tyler, Texas, plant that philosophy meant reducing the workforce by nearly two-thirds. Employees often work alone, making it difficult to establish an informal organization to support their integrity and safety; sometimes they are made to operate two machines instead of one, working 16 hours a day in highly hazardous conditions. This leads to physical and mental exhaustion; in addition, workers have to ask permission to leave the line. As one interviewee put it, "If you have a need, you just have to relieve it there at the machine." The interviewee went on to say, after apologizing for the expression, that some had urinated in their clothes.

Horrible incidents were related to interviewers. A man's arm was pulled under a belt system and was trapped; the belt scraped his arm to the bone as he cried out for two hours before anyone heard him. Another man was yanked headfirst into a machine and killed, the button to stop the machine being out of his reach by inches. The Safety Director in one plant said that *he didn't have the authority to stop the production line in the name of safety*—he had to ask the plant manager to make the decision. Workers who complained of risks or injuries or danger were fired. Some 60 percent of the maintenance workers at one plant had been injured. All were expected to lie to investigators from outside agencies.

"Disciplined management practices" at McWane translates into "it's production come hell or high water," said a design engineer who worked for McWane for 42 years. It means bright orange signs everywhere with the command: "REDUCE MAN HOURS PER TON." A former manager, who now regrets having worked for McWane, says the slogan "disciplined management practices means," and I quote, "an abusive management culture that placed the production of pipe above all else." It also meant calculating, and I'm quoting the *Times* story, "down to the penny per ton the cost of OSHA and environmental fines, along with raw materials" (4). They obviously believe it to be cheaper to pay the occasional fines than to comply with the safety rules.

What could be McWane's answer to all of this? The company refused to give an interview to the investigators, so we have little direct evidence. We do know their defense in an investigation into the death of Rolan Hoskin.

> The essence of McWane's defense—set forth in response to a civil suit brought by Mr. Hoskin's family—was that senior executives were blameless, that Mr. Hoskin and perhaps low-level supervisors had failed to abide by corporate safety procedures. (*New York Times*, 13)

That is a combination of two well-known communication transgressions: *plausible deniability* and *blaming the victim*. Blaming the victim is self-defining but plausible deniability needs some unpacking by means of a hypothetical illustration. Suppose we want to commit illegal or illicit or immoral acts and don't want to be held accountable for them. We commission other people to do the acts but make it clear that if they get caught, we want to be able to deny, plausibly, that we ordered the acts or knew about them.

We also know McWane's discursive justification of their disciplined management practice because they did respond to the investigators by letters and by e-mail. I'm now quoting from the *Frontline* transcript: "In a letter to us McWane pointed out that, in the real world, they're fighting for survival, competing against foreign manufacturers who have 'little or no regard for the safety of their workers or . . . the environment' . . ." (15). In short, their answer is—our foreign competitors made us do it. This is part of the dark side of the new global economy.

The *Frontline* program is invaluable for many reasons, one of which is that it reported the results of a naturally occurring experiment, an experiment involving a competing organization. Birmingham, Alabama, the home of McWane, has long been a city of fire, smoke, and iron. There is another foundry in Birmingham, an older one with a safe workplace and a clean environmental record in the same business of melting metal and making pipe. They've had accidents, but they represent a small fraction of McWane's numbers. Unlike McWane, they have little turnover and a long waiting list of would-be employees. The firm is consistently rated by *Fortune* as one of the best employers in America, and its workers are said to be "more productive" than McWane's. Its name is the American Cast Iron Pipe Company, and it goes by the acronym ACIPCO.

Frontline interviewed Van Richey, CEO of ACIPCO, on camera, asking him about the firm's culture and whether supervisors and workers could shut down the production line. He answered that safety is number one at his plant, that safety takes precedence over production, and that workers and supervisors are more likely to be disciplined for not shutting down the line in the face of a safety risk than for shutting it down when they are in error. John Eagan built the foundry and willed it to his workers when he died in 1924. "He was a devout Christian who tried to practice what his pastors preached by running a business based upon the Golden Rule" (*Frontline*, 17).

Frontline asked CEO Richey about this practice on camera. He said, as I quote the transcript: "Do unto others as you would have them do unto you. Before you make a rule, say, 'OK, this rule, what if it is applied to me? Is it fair?' "

The peroration of the transcript is as follows:

> There was one notable dissent when John Eagan announced that he was going to operate his foundry on the Golden Rule. The president of the company quit. His name: J. R. McWane. He crossed town to start his own pipe company and a dynasty based on a profoundly different vision. That was more than 80 years ago. In a new millennium the McWane foundries thrive, still faithful to his austere philosophy, now called disciplined management practices. They have an enviable record of commercial success and an unenviable reputation as one of the most dangerous workplaces in America. (17)

I've spent considerable time on this case because I wanted listeners and readers to know about the remarkable experiment. In addition, the dialectic between the Golden Rule and Disciplined Management Practices is one we need to consider; moreover, it illustrates perfectly two opposing types of action I shall discuss later. Until I get to the theory I shall ask you to remember that McWane and its disciplined management practices represents what Max Weber called instrumental rationality; ACIPCO, by following the Golden Rule, represents what he called value rationality. McWane's problems appear to be different from those of the other organizations and institutions I intend to discuss in this chapter, and yet their problems are similar to the findings of CAIB discussed in Chapter Seven—how to improve the safety record of an organization.

A yearly roundup in the *Denver Post* in late December of 2002 identified a long list of corporations in which one or more individuals had been accused or convicted of some kind of fraud. The list includes Enron, Lehman Brothers, Arthur Andersen, ImClone, Merrill Lynch, Adelphia, Rite Aid, Qwest, Worldcom, and Tyco. The article also said "Wall Street's biggest brokerage firms agreed to pay almost $1 billion in fines. In exchange, investigations into whether they routinely duped the investing public and handed out hot new shares to elite corporate clients will be halted" (5K). Halting the investigations will prevent us from learning more about illegal and immoral corporate actions, and some people who ought to make the "perp walk" will not have to. Nonetheless, certain changes are being made, claimed Eliot Spitzer, New York State Attorney General, to restore integrity to the financial firms and return confidence to the investor. In addition to these errant firms, the Securities and Exchange Commission (SEC) was said to have been asleep at the wheel; Harvey Pitt, its head, was forced to resign amid charges he was too cozy with people under investigation. In addition to these organizational scandals we had bankruptcy declarations by U.S. Airways, United Airlines, and K-mart, among others.

I saw Jack Welch interviewed on PBS after his small fall. Welch was probably the number one guru in the business world when he was riding high at General Electric. He lost some stature when the editor of the *Harvard Business Review* resigned after admitting she had an affair with Welch while doing an article about him. Welch's wife divorced him and in the process revealed how much he got in his Platinum Parachute at retirement; the public response to the size of his retirement package forced Welch to give some of it back. In the PBS interview he argued that there are just a handful of corporations that behaved badly, probably 12 of them, or 20 at the most. By contrast, in an interview with *Fortune,* Arthur Levitt, former SEC Chairman and referred to by the magazine as an insider and the "Sage of Wall Street," was asked:

> Q: But is this really a case of just a few bad apples spoiling it for everyone?
>
> A: No. It is a *systemic problem* represented by an almost two-decade-long erosion of *ethical values* on the part of American business (emphasis added).

Emphasis was added in the typing of that answer, emphases on "systemic problem" and "ethical values" as a forecast that we will return to those concepts later; I also emphasized those words as support for the notion that we are not talking about isolated examples or a small, skewed sample of organizations.

But there is more: *Time* magazine's Persons of the Year for 2002 were three women, three whistle-blowers—whistle-blowing, as we shall see later, was defined by W. Charles Redding as *dissent against intolerable evils directed to people external to the organization:* Cynthia Cooper of Worldcom, Coleen Rowley of the FBI, and Sherron Watkins of Enron. By honoring these women the magazine established that the FBI could be added to the list of organizations in crisis. The cover story about the women used a phrase more than once, a phrase I used as the subtitle of this chapter: *"speaking truth to power."*

In addition, we know the Catholic Church in Massachusetts and other states must be added to the list.

The American family is also a candidate in part because of the fatherhood failure; at the homeless shelter where I have worked as a volunteer for the past five years we are seeing more and more mothers with small children; 550,000 children in this country are in foster care.

Baseball, once the National Pastime, averted a total disaster by avoiding a strike in the 2002 season; nonetheless, the game lost a staggering amount of money and fans—13 percent in Denver—during that season; it is one industry where the high wages of the workers, i.e., players, are resented more by the people than the salaries of management. Ironically, the new collective bargaining agreement allows the owners to move from the raw capitalism of a free-market economy to a more socialistic scheme in which the owners, like those in football, will share revenues with each other and place limits on how much money the richest owners can pay for the top talent. A war for the top talent nearly destroyed the institution.

The Red Cross and other eleemosynary, or charitable, organizations are in trouble for failing to direct donations to causes the donors thought their contributions were going to, and for spending the money on high administrative costs. The Supreme Court, the judiciary, and the legal profession are under attack from both the Left and the Right; I recently read a new book by Omar Swartz in manuscript form, the title of which is *In Defense of Partisan Criticism,* and in which he makes a devastating criticism of the legal institution in the United States.

Agencies responsible for the regulatory-investigative function of the government seem to have failed us. I am referring to the SEC, OSHA, the FBI, and CIA. The President of the United States has appointed a bipartisan Commission to investigate the intelligence failure about weapons of mass destruction in Iraq.

Is this all a passing fancy? As I was revising this speech into a chapter, many other stories broke into the news. Three come to mind: There was an investigation into the investigations of the rape scandals at the Air Force Academy. "Ex-Leader of Tyco Wants Tapes Sealed" is the title of a small article in the Business Section of the

New York Times for Saturday, September 20, 2003. A lawyer for L. Dennis Kozlowski, the former head of Tyco International, moved to prevent a jury from seeing videotapes of a birthday party Mr. Kozlowski gave for his wife.

Mr. Kozlowski and Mark H. Swartz, the former CFO (Chief Financial Officer) of Tyco International, have pleaded not guilty to charges they stole $600,000,000 from Tyco. One of the specific charges is that Kozlowski spent $2.1 million on his wife's fortieth birthday party and charged most of it to Tyco. "Mr. Kozlowski gave the party at the Hotel Cala di Volpe in Sardinia. It featured an ice sculpture of Michelangelo's David with vodka streaming from its penis into crystal glasses." The videotapes also capture an exploding birthday cake with "a replica of a woman's breasts on top, according to people who have seen it" (B2).

Perhaps less decadent but more important was the flap surrounding the dismissal of Richard Grasso as Chairman of the New York Stock Exchange. Kurt Eichenwald of the *New York Times* wrote a long story about it, reprinted in the *Denver Post* on Sunday, September 21, 2003 (7A). He began his story this way: "Enron. WorldCom. And now the New York Stock Exchange." The point of this "lead" for his article is that the flap about the NYSE is related to the business scandals we've touched on.

'Wal-Mart Effect'

Al Lewis is a business writer for the *Denver Post* not known for left-leaning columns, but on Sunday, October 5, 2003, he wrote a column with this controversial title: "Wal-Mart has Denver in Its Vise" (1K). A controversy had been building about Wal-Mart's demand for a $10 million tax subsidy as an inducement to build a new store in a blighted shopping center. Lewis cited *BusinessWeek* magazine as raising the question as to whether Wal-Mart has become too powerful. It is now the world's largest company, selling 30 percent of America's consumable staples. Everywhere the company opens a new store it will put out of business, on average, two grocery stores plus numerous mom-and-pop enterprises. It puts the squeeze on every institution near it, including local government. It squeezes its suppliers to relocate manufacturing plants in foreign countries in order to use cheaper labor and lower their prices. The "Made in

America" slogan is long gone. Now Wal-Mart is putting the squeeze on the city of Denver.

"Economists," wrote Lewis, "in trying to explain our nation's drive for ever-greater *productivity* and efficiency, now speak of the 'Wal-Mart effect'" (1K). Although many in Denver are critical of the pressure for the huge tax subsidy, we should ask—what is wrong with productivity and efficiency? Part of the answer can be found in Lewis' column. After noting that Wal-Mart is the largest employer in Colorado as well as the United States, he added that "if you work at Wal-Mart, you probably can't afford to shop there." The average Wal-Mart clerk makes less than $14,000 a year, and in recent years its wages were below the official U.S. poverty line. Local communities often have to subsidize the company by providing costly services for its employees.

This column on the business page of our local newspaper calls to mind a brilliant and widely acclaimed criticism of U.S. business organizations: Barbara Ehrenreich's *Nickle and Dimed: On (Not) Getting By in America* (2001). Ehrenreich is a successful writer who decided to see what it would be like to join the working poor, locking up her credit cards and trying to get by in jobs at entry-level wages. She worked as a waitress, maid, house cleaner, nursing home aide, and, yes, as a Wal-Mart salesperson. The book, and the chapter about her experience working at Wal-Mart, would be depressing if it weren't simultaneously hilarious.

She makes the reader laugh at the attempts of the company to make employees identify with it. She and her co-workers making $7.00 per hour in Minnesota were called "associates" and "team members"—even though they were *forbidden to talk to each other during the workday.* Indeed, communication among workers was considered a type of "time theft." This rule makes it difficult, if not impossible, to create an informal organization to defend workers' integrity and develop a labor union. Ehrenreich observes that she would have been a better fit with Wal-Mart's culture if she had been a deaf-mute or autistic. The customers were called "guests." Her bosses were called—and I'm not making this up—"servant leaders."

"The chance to *identify* with a powerful and wealthy entity—the company or the boss—is only the carrot. There was also a stick. What surprised and offended me most," wrote Ehrenreich, "about the low-wage workplace (and yes, here all my middle-class privilege is on display) was the extent to which one is required to surrender one's basic civil rights and—what boils down to the same thing—self respect" (208, emphasis added). She reveals, for example, that she was required to take a personality inventory and a drug test as part of the application process. As she put it, "urination is a private act and it is degrading to have to perform it at the command of some powerful other" (209).

Although many consumers who identify with and are loyal to Wal-Mart think Ehrenreich overreacted, a recent poll revealed that the vast majority of Americans say they don't want a Wal-Mart store built near them. I recommend that people read the book and make up their own minds. Ehrenreich claimed that she couldn't survive on the wages from Wal-Mart, not if she wanted to sleep indoors. Housing costs have gone up rapidly in the past decade while wages have not gone up much at all because of what we will come to call rationalization, the unrelenting drive toward ever-increasing *production* and *efficiency*. In the final chapter of her book Ehrenreich quotes figures (several years out of date by now) demonstrating that a family made up of a single parent with two children needs $30,000 a year to get by on a Spartan regimen. As Al Lewis pointed out, a Wal-Mart clerk in 2003 makes less than half that figure, not enough to allow them to shop there.

Finally, more than a dozen people reviewed this book in its manuscript stage. Several of them urged me to add other organizations and whole industries to the catalogue. Among them Boeing, suggested especially for this book because of its importance as an aerospace contractor; it was nominated when Phil Condit, its CEO, or Chief Executive Officer, was forced to resign in response to allegations about two scandals that came to light in December of 2003. The first was the news that Boeing employees stole thousands of documents from its main competitor, Lockheed Martin. The company used the documents to get a contract for rocket launchers. The second scandal centered around reports that Boeing got a lucrative contract with the Air Force by offering a job to the person in charge of acquisitions (and she did take the job). The company insisted that Condit didn't know about or authorize the misdeeds, but Condit said he would resign for the good of the company. Another reviewer insisted that I add the mutual funds industry because of the scandalous discovery of late trading—letting friends buy stocks after the markets had closed. If the friends knew of information not available at the end of the official trading day, they could make a quick and sizable profit. This practice is, of course, a form of stealing from the honest mutual fund investors hoping their retirement plans would grow over the years. There were also surveys in the media about the solution adopted by the corporations in light of their scandals. They hired ethics officers called "ethicists," proof of the nature of their problems. The ethicists are reported to have considerable authority with which to police their organizations, but critics doubt they will be able to prevent future transgressions.

II

After I gave the speech about these organizations in crisis on January 18, 2003, I learned of the tragic *Columbia* breakup and thought: *Should I add NASA to the list of organizations and institutions and organizations in crisis?* I resisted doing so, partly because of old loyalties, my identification with the once-proud treasure of organizations, but after reaching this point in the narrative-analysis that is this book, I realized that NASA had some things in common with those other collectives: the substitution of instrumental-*logos* for value-*logos.*

These two terms are my adaptations of the prescient thought of the greatest sociologist in history—Max Weber (1864–1920). There has been a resurgence of interest in Weber's work inside and outside sociology. The decline of Marxism as the dominant social theory among academics has made Weber more attractive to people in cultural studies and postmodernism. In fact, it was a reading of Nicholas Gane's book, *Max Weber and Postmodern Theory: Rationalization versus Reenchantment* (2001) that brought me to a new understanding of his work. It is consistent with an interpretation of Weber as a theorist of rhetoric and communication (Tompkins 1987). I shall present Weber's thought here as a kind of social theory of rhetoric and communication.

Weber argued that ascetic Protestantism, those denominations influenced by Calvinism, helped create a culture in which capitalism could take hold and take off. The Protestant doctrine of predestination, for example, promoted the belief that possessions were symbols of salvation, symbols identifying the elect who would go to heaven. Saving money also produced the formation of capital, which, when invested, brought more wealth to the elect. Owners and workers alike had 14-hour days because they were persuaded that God expected persons to work hard and save their money. Weber argued it was the persuasive communication of these ideas that changed the economy and the culture of Europe and the United States.

In the late 1980s I showed that Weber was a rhetorician and communication theorist as much as a sociologist (Tompkins 1987). I wrote that the burning question for Weber and his time was how to understand "uncoerced obedience." That is, how can we explain why people mainly do as they are told? His answer in short was in his three types of legitimate authority, or *Herrschaft*, a word that can't satisfactorily be translated into English. It is often rendered as "domination," but I would prefer a more communicative term such as influence. Weber also developed a fourfold typology of human action: The first two are the only forms relevant to this discussion.

The first—*zweckrational*—is usually translated as instrumental-rationality; the second—*wertrational*—is value-rationality. Instrumental-rationality is

> when the end, the means, and the secondary results are all rationally taken into account and weighed. This involves rational consideration of alternative means to the end, of the relations of the end to the secondary consequences, and finally of the relative importance of different possible ends. (Weber 1978, 26)

Value-rationality is quite different because it takes as its major premises what Weber called the "ultimate values," those unquestioned value premises that control or lead to meaningful and ethical conclusions. In general, Weber thought they came from religious inspiration. My prime example, if not his, is the Golden Rule. To complete the job of translating Weber's sociology into a communicative nomenclature I shall substitute the Greek word *logos* for his word of rationality, not *logos* as the word, but *logos* as that rationality-in-communication that leads to decision making; furthermore, by using *logos* I denote a social rhetoric instead of the formal logic sometimes implied by the word rationality. Thus, henceforth I call the two forms of interaction the *instrumental-logos* and the *value-logos*.

Gane's book, a reading of Weber from the perspective of cultural and media studies, as well as postmodernism, helped me understand better Weber's prediction of the crisis of modernity, or the mess we're in today. Our problem is *rationalization*. The rational form of authority, the bureaucracy or large organization, private as well as public, naturally adopted instrumental logos over value logos. Instrumental-logos is relentless; as Weber predicted, it is taking over the world. The instrumental logos was adopted in our capitalist culture by the biggest corporations as well as the post office, the army, and even our state universities. As we marched forward under the influence of the Protestant ethic, which helped give rise to the spirit of capitalism, we worked harder, saved our money, built corporations in which we rationally calculated the means and the ends and secondary consequences of our enterprises, all of this enacted, of course, in discourse, in communication with each other. This process of rationalization necessarily led to a secularization of the space where we work as well as the discourse we enact in it. Traditional ways and emotional ties and value logos came to be considered irrational.

Value-logic or value-rationality (now I shift from Greek to English but with the reminder that it is primarily an oral logic) is driven out by the larger need to find the most efficient, cheapest, rational solutions to goals that are equally calculable—by instrumental-logic. Lawyers are needed to make sure we can do these things legally. Science is, of course, an integral part of the process of rationalization.

Let me illustrate. Recall that all decisions at ACIPCO are based on the Golden Rule. All work rules are expected to be consistent with that ultimate value. That is the pure type of value-rationality. J. R. McWane couldn't operate in such a culture so he formed his own company around the austere, even cold-blooded philosophy of "Disciplined Management Practices." Remember that McWane's companies calculated down to the last penny per ton the costs not only of raw materials but of complying with safety regulations. That calculative methodology is the pure type of instrumental logic; it is the rationalization or bureaucratization that Weber correctly feared. This instrumental-rationality, this secularization, this rationalization has produced disenchantment, a loss of meaning to workers, managers, and their families, not to mention those of us who have learned the sad story of McWane.

There is something else at work here. Recall the testimony of former workers and managers at McWane to the effect that *production was number one; it was production come hell or high water.* By contrast, at ACIPCO safety was number one because it is much more consistent with the ultimate value of the Golden Rule, by means of the value-rationality. Gregory Desilet, a philosopher of communication, helped me come to see the possibility that instrumental-rationality could seize upon or create a new ultimate value. In McWane's case their questioning and calculations of ends as well as means and of consequences drove them to establish *production* as the ultimate corporate value. Other organizations might seize or create different ultimate values. At Enron, for example, there was exuberance about satisfying greed, about individual acquisition. The hierarchy of the American Catholic Church, where one would assume a devotion to spiritual premises, seemed to accept the need to protect the church as a new ultimate value.

Football coach Vince Lombardi used to say that "winning isn't everything, it's the only thing." That became a truism in sports; it was then adopted by the business culture. Or was it the other way around? Professional football is, of course, a business. Winning has become an ultimate value, the justification for any kind of reprehensible action taken in the name of competition. This new variety of ultimate values—the Golden Rule, Production, Individual Greed, Winning is the Only Thing, and Protection of the Organization, also illustrates the "polytheistic" nature of rationalization. We are bewildered and disenchanted because there are too many gods.

I am saying that the force of business bureaucracy has produced a time and place where the ultimate values are being abandoned and in some cases replaced by new ones calculated by instrumental-rationality. Disenchantment is the condition when value-rationality and the ultimate values are reduced to instrumental-rationality. Weber's prescience, or ability to forecast, can be seen in two ways in all of this, one of them quite unexpected. The first is that, yes, managers at McWane and Enron did press ahead with instrumental-rationality to commit acts that violated the traditional ultimate values (as well as the law of the land); the second, more surprising, conclusion is that the diminishing influence of the ultimate values sometimes counteracts the effectiveness and efficiency of the business bureaucracy. As we re-enter the underside of a cycle of greed—greed being either the opposite of an ultimate value or a "new" ultimate value—the corporation, as in the case of Enron, is transformed from a rational legal instrument into an irrational illegal conspiracy.

Does any of this connect with the problems within NASA? Perhaps the reader now realizes why this chapter begins with a discussion of a sewer pipe producer, a business by the name of McWane. Recall that a design engineer at that company said "it's production come hell or high water." Recall also the description of McWane as having an "abusive management culture that placed production of pipe above all else." Production was and perhaps still is the ultimate value in McWane's Disciplined Management Practices.

Recall the hard-hitting section of the CAIB Report when it reviewed the report of the Rogers Commission:

> The Board found that dangerous aspects of NASA's 1986 culture, identified by the Rogers Commission, remained unchanged. The Space Shuttle Program had been built on compromises hammered out by the White House and NASA headquarters. As a result, NASA was transformed from a research and development agency to more of a business, with schedules, *production pressures*, deadlines, and cost and efficiency goals elevated to the level of technical innovation and safety goals. (198, emphasis added)

Production of shuttle flights on schedule sadly became the ultimate value at NASA. An R&D agency, once the pride of the country, became like a business and placed production of flights above all else. Some NASA managers admitted error; others were still in denial when they testified to CAIB. Those in denial may well have experienced Hindsight Bias, in that previous flights had been successful despite foam debris strikes. Those successes prevented them from seeing the strikes as a possible antecedent to disaster. Language was a medium of miscommunication.

It is painful to draw a comparison between McWane and NASA; although one was a business, the other became like a business; one practiced Disciplined Management Practices, the other adopted a managerial/bureaucratic culture, both of which placed production above human safety. The Director of Safety on one could not stop the production line, technical control, to save a life. The Safety culture of the other was broken, passive, and ineffective. Both have an unenviable safety record in recent years, the one having killed nine workers, the other having killed 14 astronauts and two debris searchers.

By contrast, ACIPCO has become a successful company, a good pipe company to work for; any worker or supervisor can stop the line in the name of safety, and indeed they were more likely to be disciplined for not stopping the line during an anomalous event than for stopping it. In considering best safety practices, perhaps

NASA should contemplate the case of ACIPCO, where the ultimate value is the Golden Rule. No work rules can be introduced without managers asking, would I want this rule to be applied to me? I don't mean to suggest that things are perfect at ACIPCO—nothing is—but it has been described as an effective and humane place to work.

The Golden Rule

I hurry to say this chapter is not intended to be a treatise on Christian theology—or any other kind of theology. Despite that fact, I'm aware that a neo-Weberian analysis has a kind of eerie resemblance to the complaints of fundamentalism. That is, Fundamentalist Christians think we ought to get back to the good old ways and values of our ancestors. Having said that, I will not preach any specific religion; I shall, nonetheless, explicitly endorse the Golden Rule, recognizing, however, that readers of this book aren't likely to accept such an ultimate value unless it can be justified by something other than divine commandment or because my minister told me so. Let's look at a brief history of the Golden Rule and then try to justify it in other ways.

Few realize, I think, that a variation of the Golden Rule goes back to Hillel, the Jewish commentator on the law who antedates Jesus. Hillel said: "What is hateful to you do not do to your fellow creature. That is the whole law; the rest is commentary." Jesus said, according to the New Revised Standard Version of the New Testament: "In everything do to others as you have them do to you; for this is the law and the prophets" (Matt 7:12). There is a slight difference between the two: the first is rather passive, asking us to do no harm; the second is active, asking us to do "good." Nearly every religion in world history has a version of the Golden Rule; an ancient Greek sophist or rhetorician by the name of Isocrates (436–338 BCE) taught: "Do not do unto others what angers you if done to you by others." In fact, I endorse the Golden Rule, in these or other versions, not just because it emerged from the Judaic-Christian tradition, but because I believe the Golden Rule can be pragmatically and philosophically *justified* for use as an ultimate value in human and business relations. Indeed, ACIPCO has *empirically* and *pragmatically* justified the Golden Rule by the desirable consequences of the company's application of it.

Let's try to justify it philosophically. In his book, *The Elements of Moral Philosophy,* James Rachels supplies a philosophical justification for what he calls the "primary rule of morality":

> *We ought to act so as to promote impartially the interests of everyone alike, except when individuals deserve particular responses as a result of their own past behavior.* (1993, 185, emphasis in the original)

Rachels, who also insisted on honesty as a necessity in communication-as-community, goes on to say that this primary rule combines the best elements of several philosophic traditions, including Kantian "respect for persons." The principle appears capable of producing results similar to, if not identical with, those produced by the Golden Rule. Therefore, we can accept these rules as ultimate values on the bases of (1) religious commandment or (2) pragmatic justification or (3) philosophic justification. Some people might very well accept them on all three bases. A principle of respect for persons is, of course, basic to decent human relations, productive business practices, and communication-as-community.

What are we to do about rationalization and the march of instrumental logos? James Barker, management professor at the Air Force Academy, forwarded to me in the summer of 2002 a memo from a famous management professor. The memo was generated in response to activities at the Academy of Management Convention held in Denver that summer. A panel at that convention discussed the scandals in corporate America. Were the business schools partly to blame? The panel of professors seemed to agree that courses in ethics wouldn't do any good; they had been tried after earlier cycles of greed, and were quickly crowded out by student interest in classes in finance. The eminent professor of management wrote in his memo that the problems facing organizational and management theorists were caused by the absence of a scientific base under their teachings. He expressed an analogy: physics provides a scientific base for engineering. My response is that the management professor was himself engaged in rationalization, the rationalization of human behavior. Not long after that I found myself sitting next to a professor of physics at a meal in the mountains. After I summarized the management professor's argument by analogy, the physicist responded, not pausing, with this question: "What kind of a scientific base do you need to know to tell the truth?"

III

Before introducing the role of truth telling in all of this, I must address a suggestion made by one of the reviewers of a prospectus of this book; he suggested that my arguments about corporate wrongdoing would be interpreted by some as a serious but merely cyclical phenomenon. For that reason I didn't supply the source and date of the epigraph, or opening quotation, of this chapter. Now I can say the author was the great American poet Walt Whitman; the essay is "Democratic Vistas" and was written in 1868. I invite the reader to reread the quotation and decide for herself or himself whether the problem is cyclical. It could be that the revelations of the press and the media are cyclical. Perhaps the cycle we are in now is a particularly long one.

Perhaps the communication transgressions I found while studying corporate wrongdoing are less cyclical, more timeless, than the scandals themselves. Let me address the communicative transgressions first. Reading a pile of books and articles about Enron made me wish I had the time and space to develop a thorough case study of that disaster. Briefly, the Enron executives were not creative enough to be able to pull off their great debacle without the help of one of the largest consulting companies in the world, McKinsey & Company. McKinsey employees had done an empirical study trying to see how practices varied from the average corporation to the most successful. Three of these employees wrote a book about it with the bellicose title of *The War for Talent,* apparently not realizing that such a strategy was nearly ruining baseball. They also summarized the Parable of the Talents in their preface, exploiting the double meaning of the word *talents* and erroneously saying that Luther's interpretation of it formed the basis of the Protestant work ethic.

The War for Talent and another book called *Leading the Revolution* valorized Enron and its leaders for what now appear to be disastrous practices. In retrospect they make for hilarious reading. But Enron bought the premise of a war for talent, hired Jeffrey Skilling from McKinsey itself, and put a McKinsey representative on their Board to help guarantee they followed the blueprint. The January 11–17, 2003, issue of the *Economist* has a review of two books about Enron, one of which details greed and suicide and adultery in the company. According to the review the two different books are in agreement that the fatal move was the hiring of Skilling. I'm in complete agreement, having reached that decision before reading the review. Using McKinsey's blueprint, Skilling and Enron recruited the best and brightest from the top MBA schools, whether they needed them or not, and paid them more than the market required.

Malcolm Gladwell wrote an informative article in the *New Yorker* arguing that such people—the top talent—are not necessarily good for organizations, an issue raised by reporters covering NASA after the *Columbia* accident. Skilling gave Enron's new hires vague job descriptions and stock options and bonuses with little or no direction. Why stock options? Well, it seems two assistant professors of economics trained at the University of Chicago decided that corporations were paying these young talents like bureaucrats rather than like entrepreneurs. So they thought of incentives to reward them in proportion to the growth and profits of the companies they deigned to work for.

Enter the stock option. As the corporation prospered it would attract investors in the stock market; as stock prices increased those who held stock options saw their wealth increase at the same rate. The bureaucrats were transformed into entrepreneurs, and now had a *powerful new incentive to lie.* I trust you agree with me that lying is a transgression of the lore of communication practice and a violation of an ultimate value. It is also a violation of the Golden Rule and the presupposition of an enlightened society. It may violate the most important of ultimate values: *The truth is the presupposition of faith, whether that faith is placed in science, capitalism, ideology, or religion.* In addition, let me try to justify the truth as an ultimate value philosophically. As James Rachels said, there was one important philosopher who thought we should never tell a lie—Immanuel Kant (1721–1804). No doubt today most moral philosophers would like to qualify that a bit—as in the case of lying to save a Jewish neighbor being sought by the Nazis. Kant thought not—but we should acknowledge he wrote philosophy 200 years before the Nazis. And as for Rachels, he takes this philosophic position on communication and telling the truth:

> Our ability to live together in communities depends on our capacities of communication. We talk to one another, read one another's writing, exchange information and opinions, express our desires to one another, make promises, ask and answer questions, and much more. Without these sorts of interchanges, social living

would be impossible. But in order for these interchanges to be successful, we must be able to assume that there are certain rules in
force: we must be able to rely on one another to speak honestly.
(1993, 166)

The stock options and structural pressure induced people at
Enron to break that rule, to violate that ultimate value of communication. They also failed to link the income of certain executives to
the corporation's performance because they could still make money
by cashing stocks in when they realized the stocks were egregiously
inflated. Their transgressions became more serious as they misled
the public about the firm's performance, its income and expenses,
and then persuaded their accountants to affirm their lies, doubling
the magnitude of their sins. In those cases, *there is no more organization, only relations within networks of individuals whose common interest
is increasing their individual wealth at the expense of the owners, i.e.,
stockholders, employees, and the public.* The collective-rationality of the
corporation disappeared into a new individual-rationality unre-
strained by ethics and values. There were neither internal nor independent mechanisms of control of an effective nature, such as feedback loops and truly independent audits; this constituted another
transgression of organizational communication.

We may never know the nature and extent of Enron's relationship with the U.S. government. We did learn something about the
company's relationship with the Indian government during the
1990s from a report issued by the Human Rights Watch: *The Enron
Corporation: Corporate Complicity in Human Rights Violations.* With
the backing of the U.S. government, Enron signed an agreement
with the Indian regional government in 1992 to build a $3 billion
power plant. Opposition from the Indian people was generated by
the lack of transparency—another word for openness—in the contract and with the potential for pollution. A movement against the
plant began to develop. According to the report the dissenters were
beaten by hired thugs and briefly detained in such a way as to de-
prive them of the right of freedom of speech and the right of assembly. The report said Enron was complicit in the deprivation of those
basic rights.

The role of the accounting industry in all of this is also worth
noting. One doesn't expect front-page scandals from accountants. I
had to break the habit of picturing them in green eyeshades, poring
over their columns of numbers, when I learned of their misadven-

tures. What could motivate them to certify grossly misleading reports? You know that Enron's "independent" auditing firm, Arthur Andersen, had consulting contracts as well as accounting contracts with Enron and other companies. They thus had two sources of income that would have been risked, they reasoned, if they had told the truth by way of traditional accounting standards.

The accountants also employed another questionable practice, the old *communication power play*. This is the technique of going over the head of your boss or supervisor to get your way. Arthur Levitt, the Chairman of the SEC under the Clinton administration, knew that something was wrong early in the eight years he was on the job. Since then he has written a book, *Take On the Street,* Wall Street that is, and said in his interview with *Fortune* magazine that "almost all the problems that investors are experiencing can be traced to legislators who've been motivated by campaign contributions and care nothing at all for the well-being of the individual investor." As Chairman of the SEC Levitt realized what the corporations and accountants were up to, tried to get them to change, and, after failing that, tried to get Congress to pass legislation regulating their practices. The accounting firms went over his head with campaign contributions to legislators of both parties in Congress—and it worked. At least in the short run. I wish I had more space to discuss the advantages Enron enjoyed over competitors because of what the economists call an asymmetry of information—an expression meaning one party to a transaction has more information than the other; rather than matching up buyers and sellers of energy, Enron played both roles, making the transactions opaque or cloudy rather than transparent.

The Catholic Church has provided us with another transgressive practice of communication to be placed in the same bin or category with what we have come to call plausible deniability, a technique used by McWane, and also apparently by NASA. The Church's practice has been expressed, appropriately enough, in a Latin phrase I learned in reading a sensitive account of why the unthinkable sexual abuses of children and others could go on without effective action by the hierarchy. The Latin phrase is *ignorantia affectata*, and is translated by Thomas Keneally as "a cultivated ignorance." A leader pretends or affects ignorance. This guarantees a transgression on the receiving end as well as the sending one in a rudimentary model of the communication process. The Church was also guilty of the transgressions of a cover-up and in its responses to the *Boston Globe*'s revelations. More assuring, however, is that the organization of lay Catholics who wish to reform the church by reducing the power of the hierarchy have taken the name "Voice of the Faithful." Voice, of course, is a root metaphor, an important value of participation in organizational and political communication.

Another transgression was manifested in the practice of stock "analysts" whose conflict of interest encouraged them to recommend stocks they did not believe in, stocks in fact they knew to be junk. This is simply another form of lying. Investment bankers got caught up in similar conflicts of interest. Moreover, the recognition of whistle-blowers by *Time* magazine is a singular journalistic acknowledgment of a well-established concept indicative of profound problems of organizational communication: an unwillingness to listen to subordinates. The FBI's leadership has demonstrated not only incompetence—they can't seem to get their computer systems to work—but in addition unwillingness to learn from their own employees by listening to them, thus creating the need for a whistle-blower.

Let me give a quick summary of where we've been and where we are going. I have just identified as factors implicated in organizational crises what I consider to be transgressions of the lore of communication and the ultimate value of telling the truth. Among them are (1) lying; (2) encouraging others to lie; (3) power plays which seem to be similar to political bribes; (4) failure to listen to subordinates; (5) plausible deniability; (6) *ignorantia affectata*; (7) failure to reveal conflicts of interest; (8) the fear of speaking truth to power; (9) asymmetry of information and the absence of the transparency demanded by capitalism; (10) the suppression of "voice," the freedom of speech and freedom of assembly; and (11) the absence of feedback loops to maintain control, hence the absence of organization.

IV

I want to make one more attempt to justify the Golden Rule on pragmatic grounds before discussing what we should do about these problems. I have participated in an organization that, like ACIPCO, functions on the basis of ultimate value premises, including the Golden Rule. One enters the main room of the large building to see it is filled with tables and chairs filled with people reading, talking, sleeping, playing chess. The walls are covered with paintings, including portraits of St. Francis of Assisi, Martin Luther King, Jr., and Mohandas Gandhi—the latter to remind the staff and volunteers of a commitment to nonviolence. It is the St. Francis Center, a day shelter for homeless people in Denver. It provides shelter for up to 800 people per day. We have over 400 storage spaces large enough to accommodate a 30-gallon plastic bag in which one of our guests can store his or her worldly possessions. We provide other essential services to our guests.

Their mail is delivered to us; we sort it alphabetically and hand it to them when they produce a photo ID. We make written copies for them of telephone messages that came in while they were gone. We have telephones for them to use, making us a communication center for our homeless guests, connecting them in at least a minimal way to the outside world. But we do more. The Catholic Worker Soup Kitchen serves a meal at 3:00 p.m. on Wednesdays and Fridays. We provide access to separate showers for men and women. We have clean second-hand clothing the guests can earn by performing chores such as sweeping and mopping. We have an employment office where we connect our guests with employers—yes, many of the homeless persons have jobs.

It is a cultural truism that we see our guests at their best behavior. They know they must follow the rules, consistent with the Golden Rule, and if they don't, they get "86ed," or expelled for a period of time. The rules are also there to help the homeless people gain much-needed discipline in their life. When we apply the rules, we ask if we would want them applied to us if our roles and situations were reversed. Recently we changed a couple of rules we didn't like applying, mainly because we wouldn't like having them applied to us.

When a speaker or writer slips from "I" to "we," it is a clear signal that he or she identifies with the organization or institution referred to. I'm quite aware of slipping into the "we of identification" in my discussion of the St. Francis Center. I've been a volunteer there nearly every Friday for the past five years, one of a hundred such volunteers. The experience changed my life and the life of many others. We volunteers get no financial compensation; our only incentive is a feeling of having served less fortunate persons as well as a feeling of comradeship. Statements by volunteers, as well as by guests, appear in our newsletter, *St. Francis Sun,* and they invariably speak of the deep satisfaction they experience at the shelter. Providing basic services to grateful men, women, and children who are momentarily down and out gives us not only a relief from an instrumental world, it is also a source of inspiration and renewal. I know that value-rationality works—it is beneficial to the organization, its staff, the volunteers, the guests, and the community. I cite it as another example validating the Golden Rule. In addition, I present this brief description of the St. Francis Center to be paired with the case of ACIPCO as positive examples, as models of what value logic can achieve.

What are we to do? One of the more critical studies of our current situation is Kevin Phillip's *Wealth and Democracy.* Keep in mind that Phillips is, or was, a member of the Republican Party, a Nixon strategist who now believes his party and the country have gone in the wrong direction since the 1970s. I mention his politics on the principle that reluctant testimony—testimony against one's apparent self-interest—is credible. He believes the rise of private corporate power has created such differences in income between the rich and the poor that democracy itself is in danger of disappearing in America. He believes that the corporate leaders now have enough money to buy what they want—including elections, which if it turns out to be true, will doom the institution of democracy in the United States, at least in the short run. Phillips believes that Mark Twain correctly called the 1890s the Gilded Age, and Phillips calls the 1990s the second Gilded Age. President Theodore Roosevelt saved us from the first Gilded Age by denouncing the malefactors of great wealth, saying they had the values of a pawnbroker. I can see no Teddy Roosevelt on the horizon to save us from the second, and we should not count on a political savior because the ongoing rationalization and disenchantment of the world diminishes the power of charisma, ethos, and pathos.

Phillips believes that the demise of democracy will stimulate attempts to correct the problem by radical means. He cites the demonstrations in Seattle against the World Trade Organization. Economic disparity can create radicalism; so also can war. War could unite the growing anti-war movement with those who are disenchanted by our economic and political systems. A radical reform movement, says Phillips, is not out of the question.

I see several possible solutions: Perhaps a combination of them might help. I do believe in the power of art and investigative journalism, and after writing this sentence on January 5, 2003, I watched *The Crooked E* on CBS, a television drama about Enron. And even though this specific program may not have done much to make us reflect on the dangers of instrumental-logic and the need for value-logic that employs such ultimate value premises as community and sharing, I believe the arts can help. There is a kind of tension between the arts and rationalization; in fact, the former resists and ridicules the latter. The German legend of Faust inspired two great dramas, two great works of art. According to my copy of Webster's *New Collegiate Dictionary*, the first definition of *Faustian* is perfectly appropriate for this chapter: "sacrificing spiritual values for material gain." These plays teach us the danger of selling our souls for money and power. Investigative journalism, which clings to the ultimate value of truth, even if not with an upper case "T," can also help. I believe we can benefit from the cooperative investigation into McWane and Company—and ACIPCO as a moral opposite. I also believe that criticism can help—and this is something the communication community can do something about.

In the first draft of the speech that became this chapter I suggested a course in the ethics of communication, or perhaps units in all courses. The professors of business apparently feel that hasn't worked for them. Nonetheless, I will at least share with you what I would do if I were still teaching. I would show my students a videotape of "A Dangerous Business," the *Frontline* study of McWane, and give them the transcript and articles in the *New York Times.* I would teach them the work of Weber the rhetorician. The centerpiece of the required readings would be Charles Redding's wonderful article in *Communication Education*, "Rocking Boats, Blowing Whistles, and Teaching Speech Communication" (1988, 245–58). In it Redding defines rocking boats as internal "dissent" and whistle-blowing—as indicated above—as dissent "against intolerable evils" directed to people external to the organization. Redding concentrated on the ethics of communication toward the end of his life. In 1992, just two years before he died, he delivered a paper to the Center for the Study of Ethics in Society with the title, "Ethics and the Study of Organizational Communication: When Will We Wake Up?" In the paper Redding lamented that the first two scholarly handbooks in organizational communication contained neither a chapter on nor a reference in the index to the ethics of communication. That omission did not occur in the *new handbook* published in 2001. I would also require the reading of the *Time* cover article on the courage of the three women of the year, the whistle-blowers. I believe that George Cheney's book *Values at Work: Employee Participation Meets Market Pressure at Mondragon* would be useful in teaching how difficult it is to practice value-rationality in a globalized economy. I think students should be exposed to the cycle of greed, from the first to the second Gilded Age, the S&L bailout, and other low moments in our history. I don't think a course or a unit in all courses can prevent greed in the future, but it would be worth it if a single student learned to *speak truth to power,* both as a participant in organizations and as a critic of rationalization.

This survey has shown that rationalization, the instrumental rationality striving for ever greater production and efficiency with little regard for ultimate values, has affected many organizations and institutions in our contemporary society. Communication transgressions have been part of that general pattern, from failing to listen to subordinates, to cover-ups, plausible deniability, and lying. Not enough individuals stood up to protest what they knew to be wrong; an organization cannot save itself. This is what is meant by the phrase "*Challenger* Syndrome" in the title of this chapter. Perhaps it should now be called the "*Challenger-Columbia* Syndrome," or the "*Enron* Syndrome." The final chapter returns to NASA to consider the responses to the CAIB Report and to consider the role of the individual in an organizational society.

Notes to Chapter Eight

The author would like to thank Jim Barker, Larry Browning, Greg Desilet, Cecil Franklin, Ed Jones, and Omar Swartz for making valuable suggestions about an earlier draft of this chapter.

The material on McWane and ACIPCO came from the *Frontline* program, "A Dangerous Business," broadcast on January 9, 2002, and the transcript of the program forwarded to me by Nathan Adams of the Viewers Services office of Channel 6 in Denver. The *New York Times* article of January 10, 2003, "Deaths on the Job, Slap on the Wrist," by David Barstow and Lowell Bergman, was also acquired via e-mail.

The roundup of the bad corporations appeared in the *Denver Post* of 12/29/2002, 5K. The interview with Arthur Levitt appeared in *Fortune,* October 14, 2002, 64. An important article by Malcolm Gladwell, "The Talent

Myth," was published in the *New Yorker,* July 22, 2002, 28–33. The authors of *The War for Talent* are Ed Michaels, Helen Handfield-Jones, and Beth Axelrod. McKinsey & Company holds the copyright. The Harvard Business School Press published it in 2001. *Leading the Revolution,* by Gary Hamel, was published in 2000 by the Harvard Business School Press. My source for the lack of transparency and asymmetrical information in Enron is the published Hearing before the Committee on Energy and Natural Resources: United States Senate, January 29, 2002.

The most important source for the information about the Catholic Church, including the phrase *ignorantia affectata,* is "Cold Sanctuary: How the Church Lost Its Mission," by Thomas Keneally; it appeared in the *New Yorker,* June 17 and 24, 2002.

Gane's book, *Max Weber and Postmodern Theory: Rationalization versus Re-enchantment,* was published in 2002 by Palgrave. The page reference to Weber is the first volume of *Economy and Society,* edited by Guenther Roth and Claus Wittich, the University of California Press, 1978.

Broadway Books published *Wealth and Democracy,* by Kevin Phillips, in 2002. ✦

Chapter Nine

Chicken Little, the Ostrich, and *Spiderman*

NASA was not alone in suffering serious organizational problems in the first few years of the twenty-first century. In Chapter Eight we found that the *"Challenger* Syndrome"—instrumental rationality with intense production pressures created by a management culture without much regard for ultimate values—was widespread. Organizational processes too often became communicative transgressions, or at best miscommunication. This chapter considers the reaction to the CAIB Report and the role of the individual in an age of organizational misconduct. I shall also offer root metaphors for two cultures that practice transgressions. The final section of the chapter clarifies how the film *Spiderman* relates to NASA and the two shuttle accidents.

The CAIB Report and the Press

The headlines in two of the nation's most important newspapers tell quite a bit about the reception of the CAIB Report. The *Washington Post* for August 27, 2003, the day after it was released, screamed:

REPORT BLAMES FLAWED NASA CULTURE FOR TRAGEDY

The title for a story by Rob Stein on the left front page has this title: "Miscommunication, Bungling Halted Bid for Shuttle Photos." The title for a story by Kathy Sawyer and Eric Pianin on the far right column—a spot called the "lead-all"—was quite descriptive: "In

233

Broad Indictment of Agency Practices, *Columbia* Board Says Safety Suffered." Color photographs of Ron Dittemore and Linda Ham were presented on inside pages. Ham, the mission manager, is singled out in the stories for such activities as being concerned that maneuvering the spacecraft for imaging would delay the mission schedule.

The headline for the *New York Times* of the same day, Wednesday, August 27:

REPORT ON LOSS OF SHUTTLE FOCUSES ON NASA BLUNDERS AND ISSUES SOMBER WARNING

An article by David E. Sanger has the title, "Inertia and Indecision." A second, the lead-all or overview article by John Schwartz and Matthew L. Wald has this one: "Complacency Seen." Their story continues on page A17 with the news that Sean O'Keefe thanked the board for their blueprint for change. The article also reports that one of the first moves of CAIB was "to demand that Ms. Ham and other top shuttle mission managers be removed from the front line of the investigation team, arguing that the agency would appear to be investigating itself. Ms. Ham has since been moved out of day-to-day shuttle mission management." The overview article also contains criticism of the report by several space agency watchers, including Alex Roland, a history professor at Duke University and a former NASA historian. Roland thought CAIB was recommending shuttle flights to resume too soon.

One can also learn from the editorial stances of the *Post* and the *Times.* An editorial in the *Post* published on page A24 the day after the release of the report has this title: "NASA's 'Broken Safety Culture.' " It recalls that Sean O'Keefe had warned employees of the space agency that the report would be "really ugly." The editorial writer said it lived up to the billing O'Keefe had given it; the editorial calls the report "devastating" in its "portrayal of shuttle managers as being more concerned about 'getting on with the mission than in hearing about problems.'" "Saddest of all," continued the editorial, NASA didn't learn from the "bitter lessons" of *Challenger.* The writer agreed that blame should be distributed among Congress and successive presidential administrations that failed to give the agency the money it needed to make the shuttle operational. The final paragraph begins with a quotation from President Bush: "Our journey into space will go on." It concludes with a call for serious reflection about the purpose of space exploration.

An editorial in the *New York Times* on August 27 (A22) refers to the CAIB Report as "The *Columbia* Autopsy." By describing NASA's missteps in frightful detail, said the editorial, CAIB has shown us just how daunting it will be to revive the nation's manned space program. It has signaled its belief that "management weaknesses and cultural attitudes" were the deep underlying causes for the accident. "The panel's most striking conclusion is that the space shuttles should be phased out 'as soon as possible' as the primary means for transporting humans into orbit around Earth." The shuttle is "inherently risky," continued the editorial, but not "inherently dangerous" if NASA learns to inspect and repair it in orbit and recertify its flight worthiness. The Board believes that a successor to the shuttle must be readied for flight by 2010.

> It will also be hard to change the culture and attitudes that led the National Aeronautics and Space Administration to become complacent about risks and reduce its attention to safety while struggling to meet flight schedules on constrained budgets. The space agency can presumably set up independent offices to oversee technical and safety issues relatively free of operational pressures, but as the Board members noted, no real change will occur unless people throughout the space program decide in their guts that change is important.

The final paragraph of the *Times* editorial takes notice of the Board's criticism of the national political leadership for not granting the agency the money it needs. There should be a vigorous debate about the future. "That debate ought to take place, although with budget deficits widening under the impact of huge tax cuts and the costs of rebuilding Iraq and waging war on terror, the likelihood of pursuing grandiose goals in space seems as remote as the asteroid belt."

These editorials lend support for the idea that the managerial-bureaucratic culture was *not* created in a vacuum. The White House and Congress helped create the production frenzy. So, too, did the press. Too often the press and television chided NASA for not keeping on schedule, despite the decimated ranks of employees and shrinking budgets.

About a week after the CAIB Report was released, both Admiral Gehman and Sean O'Keefe, the Administrator of NASA, showed up for hearings before a Senate Committee. Both Republicans and Democrats called for O'Keefe to find those responsible for the *Columbia* disaster and hold them accountable. A photograph accompanying an article in the *Los Angeles Times* (September 4, 2003, A14) shows Gehman standing akimbo, one hand on his hip, listening with a skeptical expression on his face to O'Keefe speaking with both hands in the air. The caption under the photograph says: "HARD QUESTIONS: Retired Navy Adm. Harold W. Gehman, Jr., left, who headed the *Columbia* Accident Investigation Board, listens to NASA administrator Sean O'Keefe before Senate testimony." The article, by Nick Anderson, has a bold title: "NASA Boss Rejects Call to Fix Blame for Shuttle: Sean O'Keefe disparages senators' demands for what he characterizes as 'a firing squad' over the loss of *Columbia*."

"Now they talk about an accident, but it was an avoidable accident," said Senator Ernest F. Hollings, Democrat of South Carolina. "You talk about failure, but it was an avoidable failure." Senator Hollings said that a captain of a Navy ship under similar circumstances would be "cashiered." Senator Sam Brownback, Republican from Kansas, asked "How do you change the culture at an institution without changing the people involved?" *Good question*, I thought, one raised in an earlier chapter.

An Insider's View

Three new publications appeared later in September; two antithetical newspaper articles and a long magazine article came to my attention. The first newspaper article is by Marcia Dunn of the Associated Press, and it appeared in the *Denver Post* on September 12, 2003; it is basically an interview with Richard Covey, former astronaut and co-chairman of the NASA task force put in place to monitor the agency's effort to return the three remaining shuttles to flight. That interview was summarized by Dunn in her lead paragraph in this way: "A former astronaut overseeing NASA's effort to resume shuttle flights said Thursday that he sees no evidence of the kind of schedule pressures that contributed to the *Columbia* disaster" (11A).

By contrast, a long article by James Glanz and John Schwartz appeared in the *New York Times* Archives. The title of the piece is "Dogged Engineer's Effort to Assess Shuttle Damage." It features

Rodney Rocha as the dogged engineer, the very person who tried to get additional imagery for the Debris Assessment Team and failed, the same engineer who wrote the e-mail message never sent, the message in which he said it bordered on the irresponsible not to get the imagery. Motivating the article was the release of other e-mail messages and interviews with participants, all of which added up to the conclusion that

> the engineers' desire for outside help in getting a look at the shuttle's wing was more intense and widespread than what was described in the Aug. 26 final report of the board investigating the Feb. 1 accident, which killed all seven astronauts aboard.

Glanz and Schwartz correctly point out that the CAIB report didn't seek to lay blame on individual NASA managers, choosing instead to focus on the technical causes and the broken safety culture. Congress, on the other hand, had opened several lines of inquiry into the accident with the intention of holding individual managers accountable. Perhaps the most telling sentence in the article, a one-sentence paragraph, is this: "In interviews with numerous engineers, most of whom have not spoken publicly until now, *the discord between NASA's engineers and managers stands out in stark relief*" (emphasis added). That sentence, of course, is the strongest possible support for the two-culture hypothesis and the cultural fence between the two groups.

Mr. Rocha emerges from the debris that is NASA as something of a hero, a man who did more than we knew to prevent the death of the astronauts. His concerns began *before* the launch, on the eve of the liftoff, when he and a group of engineers learned that a ring attaching the solid rocket boosters to the external tank had not passed a minimum strength test. Instead of halting the launch, Mr. Rocha said that the shuttle manager, Linda Ham,

> granted a temporary waiver that reduced the strength requirements, on the basis of data that the investigation board later found to be flawed. Mr. Rocha would draw on an old rocketry term—"launch fever"—to describe what had happened at the meeting.

Launch fever is an intense and infectious desire to press the button.

The launch went ahead on Thursday and the ring held, but another concern took its place—the debris strike. Rocha watched the videotape over and over on the weekend. He sent an e-mail message to Paul Shack, manager of the shuttle engineering office, suggesting that the astronauts take an EVA or spacewalk to look at the wing. Mr. Shack never responded to the message. Rocha then showed the videotape to the debris team on Tuesday afternoon, but its quality wasn't clear enough to draw conclusions. The decision to make a request for additional photos was an easy call for the engineers, who always want more data. It was in his second e-mail message that Rocha asked Mr. Shack and other managers, "Can we petition (beg) for outside agency assistance?"

Chicken Little and the Ostrich

Rocha didn't understand that Calvin Schomburg, an expert on tile but not on RCC, had been reassuring shuttle managers that the debris strike wasn't a flight-safety issue. Schomburg spoke to Linda Ham on Wednesday; she "canceled Mr. Rocha's request and two similar requests from other engineers associated with the mission," according to the investigation board. Shack informed Rocha of the decision. Rocha said he was "astonished" and sent an e-mail message asking why. He got no answer from Shack so he called him on the phone; Mr. Shack said, "I'm not going to be Chicken Little about this," as Mr. Rocha recalled. Rocha got mad.

"Chicken Little?" Mr. Rocha said he shouted back. "The program is acting like an ostrich with its head in the sand." Mr. Shack declined to comment on this exchange. The same decision was made by Mr. Schomburg and Ms. Ham. Chicken Little was the scornful metaphor used by a manager to deride engineers who cry the "sky is falling." The ostrich is an engineer's denigrating metaphor for managers who refuse to listen to warnings about serious problems. Neither bird can fly—and neither has credibility with the other.

That same day Rocha wrote the memorable memo: "In my humble technical opinion, this is the wrong (and bordering on irresponsible) answer." Rocha told Glanz and Schwartz that his finger hovered over the "send" command on his computer but didn't touch it. He showed it to another engineer, Carlisle Campbell, who said he told Rocha, "That's a significant document." Campbell also said he got madder than Rocha and together they couldn't "believe what was going on."

Rocha continued to talk to people about his concerns, two of whom informed LeRoy Cain, *Columbia*'s flight director. Cain told them he would mention it to Linda Ham. Two hours later Cain sent an e-mail message saying that the management officials he had spoken to have no interest in obtaining any more images. Cain wrote: "I consider it to be a dead issue."

It was not dead for Rocha. On Thursday, January 23, he met Mr. Schomburg, the expert on tile, and they had an argument. Schomburg's explanation amounted to hindsight bias, saying that previous, smaller debris strikes had hit the shuttle without dire consequences and that this was therefore a mere maintenance issue. Rocha countered that if the damage were severe enough, hot gases would burn through the wing during entry. Mr. Rocha recalled to the reporters that their voices rose in volume; Mr. Schomburg thrust his index finger toward Rocha and said, "Well if it's that bad, there's not a damn thing we can do about it."

On January 24, the eighth day of the mission, the Boeing group made their flawed analysis of the debris strike problem with the Crater technique; those findings were taken to an 8:00 a.m. meeting of the mission management team, "led by Ms. Ham. When a NASA engineer presented the results of the Boeing analysis and then began to discuss the lingering areas of uncertainty, Ms. Ham cut him off and the meeting moved along. The wing discussion does not even appear in the official minutes."

Another person talked to Glanz and Schwartz, a man named Dan Diggins, a CAIB investigator who had done much of the interviewing for Chapter 6 of the CAIB Report, "Decision Making at NASA." He also wrote the first draft of that chapter. Diggins told the reporters that it shouldn't be surprising that such a critical issue received such perfunctory treatment. A mission management meeting, he said, is "an official pro forma meeting to get it on the record." The real decisions had already been made. By then Rocha felt he didn't have enough data to make a sound engineering decision; nor did he have enough support to continue the fight to get the data.

On February 1, the day Gregory Williams went to see *Columbia* land at KSC, Rodney Rocha got up before dawn. He wanted to be in the mission evaluation room by 6:45, well before the shuttle fired its rockets to drop out of orbit and into the Earth's atmosphere. At 7:54 four sensors on the left wheel well went dead. "I started getting the sick feeling," said Rocha. He called his wife and asked her to say some prayers. Someone came in and said, "Oh, Rodney, we lost people, and there's probably nothing we could have done." Rocha raised his voice again, "I've been hearing that all week," he snapped. "We don't know that."

Since the accident Rocha has been thanked by engineers and his immediate supervisor. "But from management, he said: 'Silence. No talk. No reference to it. Nothing.' " One day he read an interview that Sean O'Keefe gave, in which he wondered why engineers had not raised the alarm throughout the agency's safety reporting system. "This time, Mr. Rocha broke the rules; he wrote an e-mail message directly to Mr. O'Keefe, saying he would be happy to explain what really happened." Mr. O'Keefe sent the NASA general counsel to interview Rocha and report back. After he was briefed by the counsel, O'Keefe said Rocha's experience stressed the "need to seek the dissenting viewpoint and ask, 'Are we talking ourselves into this answer?' "

The final paragraph of this important article by Glanz and Schwartz reads:

> NASA, following the board's recommendation, has reached agreements with outside agencies to take images during every flight. And 11 of the 15 top shuttle managers have been reassigned, including Ms. Ham, or have retired.

great plan—too late

A cover article about the shuttle accident appeared in the November 2003 issue of the *Atlantic Monthly*: "*Columbia*'s Last Flight: The Inside Story of the Investigation—and the Catastrophe It Laid Bare." The author, William Langewiesche, a former professional pilot and national correspondent to the *Atlantic,* was able, as the subtitle of his article claims, to gain access to CAIB's Chairman, Admiral Gehman, and other Board members. Langewiesche, for example, accompanied Gehman when he delivered the Report to NASA Administrator Sean O'Keefe, even though he wasn't allowed to enter O'Keefe's private office when the handoff was made. When they drove away from NASA Headquarters, he asked Admiral Gehman how it had been in there with O'Keefe. The answer was "Stiff. Very stiff" (85).

Although Langewiesche doesn't articulate the two-culture hypothesis, he does supply evidence that adds additional support for it. Speaking of the list of possible causes for the accident, he wrote: "But for all their willingness to explore less likely alternatives, many of NASA's managers remained stubbornly closed-minded on the subject of foam" (73). The semantogenic theory is also supported with another statement on the same page: "[I]t had become a matter of faith within NASA that foam strikes—which were a known problem—could not cause mortal damage to the shuttle." The author observes that Sean O'Keefe was badly advised to deride publicly what he called the "foamologists."

Langewiesche is highly critical of NASA management, saying it has been of late an institution whose leadership has no vision. He uses the word "bureaucratic" in referring to NASA managers and their actions and the "suppressive culture" of the space-flight program; he quotes Gehman as saying that if "you're in the engineering department, you're nobody" (82). Together these lexical choices lend support to the notion of a managerial/bureaucratic culture that defeated the engineering/team culture and prevented it from getting the imagery that could have given the agency a chance to come up with a life-saving solution.

Although Langewiesche seems to accept CAIB's decision not to scapegoat individuals, a highly unflattering portrait of shuttle managers is painted, most notably in the person of Linda Ham. He wrote that she has come to embody "NASA's arrogance and insularity"; he says she had to be "forcefully separated" from the investigation; she is pictured as "intimidating," "hard-charging," and a "youngish, attractive woman given to wearing revealing clothes" (80). *(What does her clothing have to do with her competence as a manager?)*

A more relevant and damaging criticism relates to her communication skills and is described in her answers to interview questions posed by a representative of CAIB. Langewiesche was able to reconstruct the interview. She was asked by the interviewer, "[H]ow do you seek out dissenting opinions?" Her answer was, "Well, when I hear about them . . ." The interviewer interrupted her.

"Linda, by their very nature you may not hear about them."

"Well," she responded, "when somebody comes forward and tells me about them."

His follow-up question: "But Linda, what techniques do you use to get them?"

The interviewer told Langewiesche she had no answer (p. 82).

No earthquake prediction system here, as well as inadequate superior-subordinate communication skills, I wrote in the margin of the page in the *Atlantic*.

Langewiesche also gives communication attention by reporting the views of others. One is Edward Tufte, "the brilliant communications specialist from Yale." On the ninth day of *Columbia's* flight, engineers who were part of the Debris Assessment Team gave a briefing to Linda Ham's "intermediary," Don McCormack. They projected a crude PowerPoint summary of the findings of the Crater model. They concluded that if the debris hit the tile, *Columbia* would probably be OK, but they also tried to communicate their uncertainty. Unfortunately their uncertainty was filtered out in such a way that Linda Ham heard, by relay, that there was no safety issue. Tufte issued a booklet, *The Cognitive Style of PowerPoint,* in which he argued that the method has a negative effect on clear thinking and clear speech. CAIB agreed and cited the widespread use of PowerPoint in NASA as a serious obstacle to internal communication.

In addition, Admiral Gehman had a lot to say to Langewiesche about communication. NASA talked the talk and said all the right things about communication, Gehman argued; they say, yeah, we've got open doors and e-mail and people can blow the whistle and speak up.

> But then when you look at how it really works, it's an incestuous, hierarchical system, with invisible rankings and a very strict informal chain of command. They all know that. So even though they've got all the trappings of communication, you don't actually *find* communication. It's very complex. But if a person brings an issue up, what caste he's in makes all the difference. Now, again, NASA will deny this, but if you talk to people, if you really listen to people, all the time you hear "Well, I was afraid to speak up." (82)

The expression "caste" makes it clear that the engineering culture was the lower class; the management/bureaucratic culture the upper class. And the upper class is ruthless. Gehman quoted witnesses who said "If I had spoken up, it would have been at the cost of my job." Again we have support for the two-culture hypothesis; the burden of proof has been reversed in a general way. People feel they will be punished for speaking up. There were no Monday Notes, no automatic responsibility, and the open communication has been changed to discursive closure.

Langewiesche thought Sean O'Keefe had resisted CAIB's attempts to be truly independent. He also wrote that Admiral Gehman's biggest mistake—or "arguably" so—was the decision to allow witnesses interviewed by the staff to be "protected." They were allowed to speak to the interviewers on a confidential basis so as to encourage them to be completely open. As I read that criticism, I disagreed with Langewiesche's evaluation. I thought back to my own interviews with NASA personnel at three different data points in time. Never did I attribute an answer to a specific person. I had no doubt then and have no doubt now that this encouraged them to open up to me.

Langewiesche's article, as mentioned above, appeared in the November issue of the *Atlantic Monthly*. The January/February (2004) issue of the magazine contained a flurry of letters to the editor in response to the article. The first was by Sean O'Keefe, who complained that he had cooperated with CAIB. He also said NASA is taking "steps to develop an organizational culture that empowers open dialogue and rewards excellence in all aspects of our work" (p. 12). O'Keefe also said he was "disturbed" about Langewiesche's claim that no one "stepped forward to accept personal responsibility for contributing to this accident." O'Keefe said he had done precisely that in a March session with reporters and will continue to act in that manner for the rest of his tenure at NASA.

A second, long letter was written by John G. Stewart, member of NASA's Aerospace Safety Advisory Panel (ASAP) from 1980 to 2001. His main argument is that Langewiesche failed to place sufficient blame on the White House, the Office of Management and Budget (OMB), and Congress for "funding shortfalls and personnel reductions." He also provided another potential problem of language and meaning: "Since its earliest flight the space shuttle has been an R&D vehicle, despite sporadic attempts by NASA and OMB to declare it 'operational.' The distinction is more than semantic. The space shuttle is the most complex machine ever conceived, built, and flown by human beings" (12). "It must never be seen as 'operational,' " continued Stewart, "in the sense that a Boeing 747 is operational" (16). By calling it "operational" instead of a Research & Development project, NASA and OMB applied a label that caused people to behave differently around the shuttle than they should have. The words *operational* and *foam* affected human behavior in a disastrous way.

"William Langewiesche is dead wrong when he states as 'fact' that allowing witnesses to be protected in *Columbia* Accident Investigation Board interviews was arguably Gehman's biggest mistake. . . ." So wrote Duane W. Deal, USAF Brigadier General and a member of CAIB.

> As one who conducted a third of the interviews, I saw firsthand that we gained valuable information—information we likely would not have gained if the comments had been open to public scrutiny. Indeed, not unlike reporters protecting their sources, we had people offering testimony under privilege that could have cost them their jobs if made public. Such testimony, particularly when validated through multiple interviews, went on to help us piece together the dysfunctional communications within NASA. . . . (16)

Another letter writer said Langewiesche should have addressed the issue of the nature of the insulating foam—that the original foam was made with Freon, which was banned by the Environmental Protection Agency. NASA, according to this letter, substituted foam without Freon, which was inferior. Still another writer defended PowerPoint; this "much-maligned Microsoft application is only as good as the message developed by the user" and that relying on arguments against it is "to revert to the custom of blaming bad news on the messenger" (16).

A reader named Lori McKay sent this succinct letter:

> I was surprised by William Langewiesche's description of Linda Ham. He writes that she was a "youngish, attractive woman given to wearing revealing clothes."
>
> The *Columbia* flight did not end in tragedy because some engineer, rather than minding the computer readouts, was busy looking down Ham's blouse. I fail to understand why her dress code was included. One can make the point that she was a successful professional woman in a male-dominated industry without reference to fashion tastes. Ham committed egregious errors in management decision-making. To comment on what she was wearing when she made those errors detracts from the very real thrust of the article: seven people died because the organization charged with protecting them had an abysmal failure of communication at all levels. (16/18) (*Bravo.*)

Finally, a reader of the Langewiesche article wrote that he was "startled" to read in it that when astronauts return to the shuttle from a space walk their gear has a strong space odor, an acrid or burned quality. "Space smells? Where does the odor originate?"

Langewiesche replied that neither he nor the astronauts he asked understand the origin of the smell. I asked Gregory Williams, an engineer, about its origin.

"Ozone," was his one-word answer.

The Meaning of the Situation as Influencing Behavior

In Chapter Three I presented a definition of organizational culture that stressed meaning. In retrospect we can see how this central dimension of organizational culture manifested itself in a critical way. One exhibit is the now-apparent confusion within NASA about the meaning of the impact of debris on the orbiter. Most of the time NASA managers and engineers concentrated on the tile on the wing and around the wheel well. The Crater analysis made by the Boeing employees assumed the "acreage," or area of the shuttle "hit" by the foam, was covered with tile. Occasionally the reporters would bring up the RCC, teach their readers about its nature and function, only to have NASA briefers return to the subject of tile. In the end, of course, it was the RCC that was breached by the foam debris, not the tiles.

CAIB discussed a second problem of meaning. The shuttle managers didn't seem to understand the terms, or categories, used to characterize the degree of risk associated with an anomaly. I would cite "in-family" as the most appropriate term that affected behavior. The two-word phrase is applied to an anomaly that is perceived as similar to another problem they understand well. At times it turns out they don't understand it all, but the phrase "in-family" makes them behave unthinkingly as if they do. A third problem of meaning—the meaning of the word *foam*—was perhaps the most important: Everybody knows that *foam* can't hurt a shuttle—even if it is as hard as a brick. Recall that Sean O'Keefe told a reporter that the debris strike would be like a Styrofoam cooler being blown out of a pickup truck and striking your car at 50 miles per hour. Recall also that some NASA personnel predicted that the foam fired at an RCC panel during the test in Texas would bounce off like a Nerf ball! No doubt some underestimated the danger by being misled by the expression "debris strike" or a piece of foam "hitting" the orbiter or the foam debris "falling off" the External Tank. These expressions are more passive, less threatening, than saying the *Columbia* "ran into" the debris at a relative velocity of 545 miles per hour—not hitting a Styrofoam cooler at 50 miles per hour.

We considered earlier Benjamin Lee Whorf's theory of "The Name of the Situation as Affecting Behavior." I debated changing the word from *behavior* to *action* or *choice* because, as Kenneth Burke taught, humans literally act. Another reason is that the behaviorist theory of psychology holds that we simply respond to stimuli: Behavior is the result of the stimulus-response pair. I decided against the change because there is something behavioristic, or unthinking, about how people responded to "empty" gas drums, "limestone," "in-family" problems, an "operational" space shuttle, and "foam strikes." Production pressures no doubt make management react in an unthinking way. Some engineering teams did think about what foam might do when hit at over 500 miles per hour. Sadly, the two groups could not communicate with each other in such a way as to understand the difference and try to bridge the different cultural meanings.

Ideal Managerial Climate

In Chapter Three I also introduced Redding's Ideal Managerial Climate. After looking at all the evidence about NASA it is appropriate to apply that standard to the evidence. It would also be instructive to keep McWane, Enron, Wal-Mart, and other such organizations in mind while applying the five descriptors: (1) *Supportiveness.* Subordinates were not supported by management. Managers supported each other but not those down the hierarchy: Negative. (2) *Participative Decision Making.* Three different engineering groups worked together to solve the foam debris problem. Their requests for additional data were denied. At Wal-Mart employees were forbidden to talk to each other; communication was time-theft. At McWane the employees worked alone; no one could hear the screams of the man whose arm was caught in a machine: Negative. (3) *Trust, Confidence, and Credibility.* NASA engineers feared they would lose their jobs if they told the truth. Communication transgressions at other organizations destroyed management's credibility: Negative. (4). *Openness and Candor.* No discussion required under this point: Negative. (5) *Emphasis on High Performance Goals.* In today's spirit of rationalization, performance and production goals are high, so high that they can displace the ultimate values of the

Golden Rule, telling the truth, and the sanctity of life: Positive. One for five is not a good average; and the one positive turns out to be a negative.

To Scapegoat or Not

In Chapter Six of this book I presented an antinomy, two contradictory truth claims, both of which can't be true. The one truth is the admonition against scapegoating; the other is the desire to fix responsibility, to hold individuals accountable. My conclusion is that one can only do this on a case-by-case method or casuistry. I developed eight criteria for resolving an antinomy. I propose now to apply the list of criteria to the *Columbia* case.

1. *Causal force.* Did the action or inaction of an individual have causal force or did it amount merely to being one in a long chain of causal events? My answer is that certain NASA managers did by action and inaction prevent additional imagery from being obtained so that engineers could assess the damage and work on a way to prevent the loss of life.

2. *Hierarchy.* What was the person's formal degree of authority and responsibility? CAIB found that the shuttle managers were located at a high and central position in the communication structure and did act in ways that stifled and squelched the concerted efforts of engineers.

3. *Values of the culture.* Was the individual's act consistent with the ultimate and avowed values of the culture? No. The shuttle managers showed a lack of concern about the lives of the astronauts and about the facilitation of dissent, upward-directed communication from the engineers. Despite the broken culture, or development of a new managerial/bureaucratic culture, the ultimate value of manned flight was and is safety, a reverence for life.

4. *Consequences.* Were the consequences of the individual's actions or inactions trivial or significant? Significant. Seven lives were lost.

5. *Justice.* Does the punishment fit the crime? Would the effect of blame or censure be commensurate with the act? Yes. I'm not proposing a firing squad or electric chair or gas chamber

or even a jail sentence. It does seem appropriate to remove responsible persons from their positions and dismiss them from the agency.

6. *Defense.* Did the person accept responsibility or truthfully deny responsibility? Ron Dittemore said immediately he was the "accountable individual" and ultimately resigned from the agency. CAIB concluded that others were in "denial" about their responsibility for the accident. Administrator O'Keefe said in his letter to the editor of the *Atlantic* that he felt a "deep sense of personal responsibility."

7. *Actor agency.* Did the person have autonomy, or control over the act or attribute that had causal force? There is no evidence in the CAIB Report to suggest that the shuttle managers were unable in any way to call for additional imagery and thereby delay the mission schedule and subsequent flights.

8. *Future actions.* Might the person accountable make the same mistake again? This is the most difficult of my eight criteria to apply. It calls to mind von Braun's rule—forgive a person for the first mistake and he or she will never make that mistake again. I was once interviewing an MSFC manager in his office when a phone call came in from von Braun. He took the call and I heard his side of it and we discussed it when he hung up. The Director was inquiring about a problem and the manager admitted it "got away from him." Von Braun forgave him for a major mistake. But the mistake didn't result in the loss of life; so I think this criterion must be weighed against the consequences criterion. Tipping the scales ultimately for me is the fact that I can't conceive of fixing the managerial/bureaucratic culture without replacing members of management.

The application of the criteria to the evidence in the CAIB Report teaches that NASA should move to hold individuals accountable—with the possible exception of the eighth criterion. And that criterion is weakened by the serious consequences of the faulty decisions and inappropriate attitudes and premises underlying them. In addition, the need to create a new culture points in the direction of making personnel changes—by firing responsible individuals.

On the first anniversary of the *Columbia* accident, February 1, 2004—Super Bowl Sunday—there were memorial ceremonies around the country. My local newspaper, the *Denver Post,* published an article that day with this headline: "NASA STAR QUICKLY FELL FROM GRACE: Linda Ham's Future Remains Uncertain" (4A). For the previous two months she had been working just a few miles west of Denver, "on loan" by NASA to the National Renewable Energy Laboratory in Golden, Colorado. The article was written by Michael Cabbage of the *Orlando Sentinel* and reprinted in the *Post.*

"If people say there are problems with the NASA culture, I will admit that I am part of it," Ham said in an interview. "There isn't a day that goes by that I don't think about *Columbia* and the accident." She regrets saying, "It's not really much of a factor in the flight because there is not much we can do about it." She was described as speaking softly when she said, "I am accountable for any decision made at the MMT." Her assignment at the Golden Laboratory ends next summer; beyond that Ham has no concrete plans. " 'Usually,' she says wistfully, 'I know where I'm going" (4A).

The Two-Cultures Hypothesis

Satisfied with the results of the test conducted in Chapter Seven in which the hypothesis was applied to a representative sample of statements from the CAIB Report, I wish now to present it as a definitive explanation of the organizational cause of the accident. The substance of the argument is this: There is, or there was, a dangerous hierarchy in NASA in which the managerial/bureaucratic culture dominated the engineering/concertive culture. That judgment is not weakened in any way by the reaction of the press to the CAIB Report. Both the *Washington Post* and the *New York Times* agreed with the assessment that something was wrong with the NASA culture, and in assessing blame mentioned members of the shuttle management.

Although I think a "broken culture" is a bad, brittle metaphor to use in regard to an invisible but powerful phenomenon, it is necessary to ask how to fix it—or fix the problem of two antagonistic subcultures. The only conceivable solution that has occurred to me is to implement the conclusion dictated by the application of the criteria and the two-culture hypothesis: Replace much of the current shuttle management team with people whose loyalty has been to the engineering/concertive culture. "Reluctant managers" would be preferable to bureaucratic managers who stifle communication about urgent matters.

The Historical Perspective

As mentioned in Chapter Four, a historian by the name of Stephen B. Johnson, produced a book in 2002 with this title: *The Secret of Apollo: Systems Management in American and European Space Programs.* The purpose of his book is to explain the success of Apollo, its secret of success, and what the Europeans learned from it. As the subtitle of the book reveals, Johnson argues that the secret was what he calls "systems management." The short version of his definition of systems management is this:

> Systems management . . . was (and is) a mélange of techniques representing the interests of each contributing group [military officers, scientists, engineers, and managers]. We can define systems management as a *set of organizational structures and processes to rapidly produce a novel but dependable technological artifact within a predictable budget.* In this definition, each group appears. Military officers demanded rapid progress. Scientists desired novelty. Engineers wanted a dependable product. Managers sought predictable costs. Only through successful collaboration could these goals be attained. To succeed in the Cold War missile and space race, systems management would also have to encompass techniques that could meet the extreme requirements of rocketry and space flight. (2002, 17, emphasis in the original)

Johnson identifies the mixture of organizational structures and techniques making up systems management as matrix organizations, project management, configuration management, systems engineering, and component tracking and control systems. He gives credit to the Air Force for developing systems management. It was transported to NASA in large part by an Air Force officer, Samuel Phillips, who became the Apollo Program Director in Washington, D.C. On page 136 of Johnson's book there is a photograph of George Mueller and General Phillips. The caption under the photograph says that they "imposed air force management methods on Apollo by introducing new procedures and bringing dozens of air force officers into NASA's manned space flight programs."

Johnson is aware that there was much resistance to their attempts to impose Air Force methods on NASA, particularly at MSFC. He referenced, for example, my research into the Saturn V Control Center. As pointed out in Chapter Four of this book, MSFC relied much more heavily on the *informal* communication activities in the "mini control room" across the hall, a beehive of activity, than it did on the formal apparatus of the Control Center. And although MSFC used project management, the biggest problem I found in 1967 was in horizontal communication—I called it the "lack of lateral openness"—between the project offices in IO and the laboratories in RDO. Another serious communication problem I discovered was the "formalism-impersonality syndrome," which implicated the techniques of systems management.

In addition, Johnson describes my activities in helping reorganize MSFC in 1968, one major purpose of which was to strengthen systems engineering, an integral part of systems management. Recognizing that the dismissal of systems engineering and systems management by MSFC weakens his thesis that this was the "secret" of the Apollo Program, Johnson is forced to raise the question as to why MSFC, with NASA's most experienced engineers and managers, was so slow to embrace systems engineering. His answer:

> Simply put, when each team member knew the job through decades of experience and knew every other team member over that period, formal methods to communicate or coordinate were redundant. Rocket team members knew their jobs, and each other, intimately. They understood what information their colleagues needed, and when. When they began to work on new products such as space stations and spacecraft in the late 1960s, it was no longer obvious how each team member should communicate with everyone else. Formal task planning, coordination, and communication became a necessity, and systems engineers performed these new ways. (2002, 151–52)

His answer, again simply put, is that effective informal communication was part of the secret of success at MSFC during the Apollo Program.

Johnson is also correct in arguing that during the post-Apollo program of space stations and spacecraft such as the shuttle, MSFC did adopt systems management. Given MSFC's responsibility for the shuttles, how can he explain the two shuttle failures? His book was published before the *Columbia* accident, but he does discuss the *Challenger* accident in this way:

> The *Challenger* accident occurred during operations, not during R&D. Systems management, an R&D management scheme, perhaps should never have been applied for operations, but it was the only system NASA knew. Systems management contributed to this disaster not because of bureaucracy per se but because through systems management, *managers gained control over engineers.* Managers then overruled engineering concerns to keep an operational schedule. On Hubble, to save costs, managers eliminated tests that would have revealed Hubble's myopia. Is "the system," namely the bureaucratic procedures, to blame for these failings? As a communication and control system, systems management revealed the problem on *Challenger* and [system management] was reduced on Hubble so that it could not find the problem. For both projects, the issues are more complex than a simple "systems management is or is not to blame" stance (2002, 275–76, emphasis added).

This paragraph—a footnote in a book about the success of Apollo—needs to be unpacked. First, Johnson says that the *Challenger* accident was a problem of operations, not research and development. Systems management, he says, shouldn't perhaps have been used for operations. This reminds us of the semantogenic controversy over whether the shuttle should have been *named* an operational or a research and development project. Second, this explanation suggests that by adopting systems management, "the only system NASA knew," NASA was left without appropriate methods needed for manned flight operations. Third, Johnson's assertion that systems management allowed managers to gain *control* over engineers anticipates to some degree my claim that the bureaucratic-managerial culture defeated the concertive-engineering culture during the crises that doomed *Challenger* and *Columbia*.

In general we can say that Johnson's thesis about the success of Apollo applies more to other field centers than to MSFC. In addition, it is significant that Johnson finds that the formal methods of systems management were *not* developed for operations, leaving NASA without appropriate methods for the management and communication of them. Again it is significant that although Johnson did not employ a cultural analysis, he did identify "four social groups" with conflicting interests: scientists, engineers, military officers, and, of course, managers. Finally, it is clear that in finding the optimal methods of managing the research and development of complex systems in the future, as well as managing their operations, we will have to cope with the problem of how formal and informal control/communication systems interact.

Final Recommendation

On January 15, 2004, President George W. Bush gave a speech that included his vision for the future of the space program: "Continue the journey." Recent presidents have been faulted for not paying enough attention to NASA and not providing leadership for the space program. No doubt motivated by those criticisms and by the shock of the *Columbia* accident, Bush laid out his plans in a televised speech. The basics of the plan are: (1) Return the space shuttle to flight and fulfill our commitments to the International Space Station; (2) halt work on the ISS and retire the shuttle fleet by 2010; (3) use robotic explorers throughout the solar system; (4) develop a new "crew exploration vehicle" for venturing beyond Earth's orbit, test it in 2008, and launch it for the first mission by 2014; and (5) return astronauts to the Moon between 2015 and 2020, establishing a long-term base there for manned missions to Mars. It is a bold vision. The basic question about it is, can we do it with our current and projected budget deficits?

I would endorse such a plan if the White House and the Congress provide the budget support and new personnel to strengthen the engineers and managers and tear down the cultural fence between them. I would create a new rule: the 15-percent rule for manned space flight. Recall that during the successful Apollo Program, von Braun insisted on having at least 10 percent of the workforce as employees of NASA, allowing the contractors 90 percent. The 15-percent rule would go beyond that and could help achieve several objectives. First, it would revive the notion of the Arsenal, particularly if the ratio of managers to engineers among the 15 percent was the same as or close to what it had been during the Apollo Program. It would also allow the reintroduction of the Penetration Principle by assuring that the in-house technical capability necessary to monitor the work of the contractors was at hand. In addition, it would strengthen the engineering/concertive culture in relation to the managerial/bureaucratic culture. This, plus the independent safety and technical systems recommended by CAIB, should allow the agency to regain its in-house capability.

I would also like to take the occasion to remind CAIB and the White House that there is still a Faustian Bargain. In fact, there is more than one of them. Faust is a German legend that was made into literature by two great artists, Christopher Marlowe and Johann Wolfgang Goethe. Their dramas about the legend are so different that they are difficult to compare. Nonetheless, we do get the adjective "Faustian" from their works and its current dictionary meaning: *sacrificing ethics for material gain.* As we saw in Chapter Eight that meaning of the Faustian Bargain applies today to many American organizations and institutions; that contract must be destroyed. The other Faustian Bargain is the one mentioned in the CAIB Report: the deal the American people have made with nuclear scientists and engineers as well as with NASA, the essence of which is that we will let you work with dangerous forces if you promise to make safety the highest, ultimate value. That bargain must be kept.

Despite the system, despite the power of the management and culture of NASA, the accident shouldn't have happened. It was avoidable. To repeat a motif if not a theme of this book, individuals failed as well as the system. What lesson can be found in all of this for the college and university student reading this book? I assume every single one of them will find herself or himself working in and dependent on an organization in the near future; that organization will be tempted to adopt the extreme form of rationalization, instrumental-rationality, in which ultimate values are ignored if not violated. I must introduce a new concept that has advice for those who face a career in an organization or series of organizations.

Ontological Anxiety

Ontological is a 25-dollar word and *ontological anxiety* is a mouthful of syllables; even though at first glance the phrase appears forbidding, perhaps impossible to understand, it can be grasped if one has patience. The phrase was coined by a philosopher-theologian of the twentieth century by the name of Paul Tillich. Ontological is the adjective for the noun ontology, the study of essence or being. Anxiety is a painful or uneasy state of mind. Tillich meant by ontological anxiety the pain or uneasiness about one's very being (Tillich, 1952/2000).

Why should a person be uneasy about being? Tillich's answer is that each of us faces a tension between our *individuality* and our *participation* in society and its institutions and organizations. As an individual we have freedom and independence, but to develop to our full potential we have to participate in social units such as the family, school, and organizations. For example, learning language is a communal enterprise. But as the individual participates in society, the institutions and organizations insist on controlling and limiting him or her—a threat to the individual's freedom, autonomy, identity, and integrity.

Tillich uses the expression "we-self" to refer to the changes produced in us by participation. I would amend that to "we-selves" because a person takes on a new identity in each of the organizations in which he or she participates. A student has a we-self in the classroom that is different from the one in church, at a keg party, in an employment interview, on a date, and on a sports team; that student will behave differently in each situation. Does this make a person less authentic as an I-self? Clearly there is a tension between the I-self and the we-self.

The resolution of this tension can't be made by taking an extreme stand on either of the two poles of the continuum. Withdrawing from organizations and society will produce a stunted, undeveloped personality that is completely self-centered. Allowing each of one's focal organizations to mold completely the we-selves would mean a complete loss of identity and authenticity. The resolution has to be sought in finding a balance between the two tension-producing poles. Loyalty to one's ultimate values will be necessary to self-respect and self-acceptance.

When an organization asks the individual to compromise his or her ultimate values—and those of the organization—resistance must be considered. It would be wise in such situations to consult the informal organization; do others perceive the situation in the same way? If they do, seek support for your resistance to the organization. If not, then it will be necessary to make a painful decision—probably with ontological anxiety—about whether to comply or not. To comply or not will depend on whether one feels that ultimate values might be compromised. In some cases, the choice will be between rocking the boat, blowing a whistle, or a quiet resignation from the organization.

It will take knowledge and ethics to know when management and the organization have gone too far. It will take courage to stand up and complain about the transgressions. A high communicative value is: Speaking truth to power.

Coda

Tom Jensen is an orchestra conductor, a radio personality, friend, neighbor, and a substitute teacher in the Denver Public School System. He wrote the following story for this book. As a professor I find the story an act of teaching genius. I was also inspired by the learning experience he describes. I hope that others will have inspired moments of learning while reading and discussing this book.

The *Spiderman*/NASA Lesson Plan
By Tom Jensen

The call came at five a.m. A secretary at a middle school in the worst part of Denver asked if I would take an assignment teaching science. She said that I would be doing her a favor as this was a tough class. Since I knew this lady from previous requests, and because she was up front about it being a hard group of kids, I agreed and got into high gear as I was to be teaching in less than three hours.

As I shaved and combed my hair, my mind was racing—I was told that there had been seven substitutes over the past month and a half since the school year started, that there was no grade book, there was no seating chart . . . and that the school wasn't quite sure exactly how many students were even enrolled in the course.

The school looks like a prison. Panopticon views and an absence of windows enabled faculty and security to better monitor the inmates. When I asked a regular teacher about the environment, he informed me that the architect who designed the school also designed prisons—in my mind getting kids ready for their "next home." Not a pretty place to learn.

Many of the classrooms have been divided up into smaller spaces because of a burgeoning population. The smaller rooms have thin, cardboard-like walls that invite cacophony when noise permeates from room to room. This was going to be eight hours of hell.

On my way out the door I saw Dr. Phil's book, *Organizational Communication Imperatives: Lessons of the Space Program* (1993). For some reason I picked up the book and threw it into my briefcase as I rushed out the door. I am not sure why I took it, but it saved my classes that day.

Driving to school, my mind still going a mile a minute, I started tap dancing through a lesson plan that might entertain inner city seventh graders, and maybe even teach something. Oh, by the way, I have no background in science, but I had just finished reading Dr. Tompkins' book.

I arrived at 7:20; class was to begin in ten minutes and there was no key to the room. A custodian was called, and with the kids all clamoring around the door we finally piled into the room about two minutes past the bell. Decorum was nonexistent and I decided to forget about taking attendance, launching into my oral explanation of my version of the space program. (There were xeroxed sheets on the desk. Evidently, for the last few weeks the kids were given a "science story" with a fill-in-the-blanks Q&A on the flip side. Every day they were to read the story and write their answers in the blanks with bold printed words from the story that answered the questions—inspired education.)

I lowered my voice and asked, "How many of you have seen the movie *Spiderman*?" It had been released a year earlier and had been a commercial hit. Most of the 35-plus students—many of whom were wannabe gang members—raised their hands. And then I did a half-gainer to the blackboard, writing: "What do the movie *Spiderman* and the space shuttle accidents have in common?"

The students started talking and yelling and I had to calm them down again—remember they had had over a half-dozen teachers in as many weeks and were bored, frustrated, and, quite frankly, pissed off.

Somehow I got them to stop talking and write down their impressions of the movie, guiding them to the part that I wanted to highlight: Why did Norman Osborn do what he thought he had to do in the film?

They started screaming about bits and pieces of the flick, but I reminded them that this was a written assignment and the only impressions they could share with the class would be the ones that they wrote down on paper.

Stan Lee, the creator of Marvel Comics, was a consultant to the movie, and the plot revolves around Peter Parker, his love for Mary Jane, and his nemesis, Norman Osborn, a.k.a. "the Green Goblin."

I am still passing out pencils and paper well into the first 15 minutes of this nightmare when a Hispanic male in the front row raises his hand and shoves his paper into my face. He has scribbled down the facts that Norman Osborn took a serum and became the bad guy. Finally the others begin to realize that they get to share their thoughts only if those ideas are written down—and a quiet ambiance engulfs the room for the first time since the beginning of the school year. (I know this because a regular teacher in the room adjacent to mine later told me so.)

Now kids are raising their hands and upon inspection of their papers we start to realize that Norman Osborn not only takes a potion but gets his lab assistant to strap him into a Frankenstein-like chair incased in a glass chamber.

OK, enough of the plot. The kids are missing the big picture, the segue that I am trying to build to make the quantum leap justifying the *Spiderman*/NASA lesson plan. But that is fine by me as the class period is not one, but two 47-minute periods without a break—a "Block Period."

An African-American male who spent a good five minutes at the pencil sharpener with his pants a-sagging and his underpants a-showing, finally wrote what I was looking for: the United States Government had been pressuring Norman Osborn to conclude his project or the funding for the mission would be withdrawn. Of course the student didn't quite express it that way but he got the basic drift.

Now I was on fire! I scribbled a quick outline on the blackboard: Adolph Hitler, World War II, Wernher von Braun and the invention of the V-2 rocket, Monday Notes, Automatic Responsibility, 1969 and the landing on the Moon, and finally the shuttle disaster. The students were told that they were to write all of this as an outline with notes from my impromptu lecture; it was due at the end of the first 47-minute period.

Then I brandished Dr. Phil's book and said: "A friend and colleague of mine has written this incredible book—a book with a very boring title! *And this book will explain what the* Spiderman *movie and the space program have in common."*

Some knew who Hitler was. A couple of kids knew about WWII. Nobody had heard of von Braun and while some had an awareness of Kennedy, they had no idea where his place was in history—a lot of this was like a bad Jay Walking segment on Leno's *Tonight Show.*

So I started with some history: von Braun leaving Germany. Our winning the war. *Sputnik* and the space race. (I didn't have the energy to get into the whole Soviet Communist thing.)

But I did get into the Monday Notes and how low-level engineers and scientists could give their thoughts and ideas and problems straight to von Braun. And Automatic Responsibility and the other communication practices that allowed the U.S. Space Program to put a man on the Moon within the ten years President Kennedy had established as a national goal.

But then I told them that the Monday Notes were changed and Automatic Responsibility was forgotten and communication broke down. And that not one, but two preventable shuttle disasters had taken place: *Challenger* in 1986 and *Columbia* in 2003.

The kids really liked getting into the dismembered body parts being scattered over Texas and Louisiana. But they also learned about "O" rings and temperature; and foam debris hitting the wing of the shuttle during liftoff. All of this because the government had tight schedules for the missions. To their credit, some knew about the foam problems and remembered watching the news of the event.

We talked about Ronald Reagan and the science teacher who died. We discussed how the astronauts on *Columbia* could have possibly been rescued.

As the 90-minute-plus epic class came to a close, a Vietnamese student raised his hand and asked the question that made my day: "Does this mean that social studies can influence what goes on in science?"

Wow! Here is a seventh grader, in a dysfunctional classroom, in a crummy inner-city school, who summarizes Dr. Phil's book with the boring title—and his next one—in a 13-word question.

It was a magical moment.

"Houston, we have a successful lift-off." ✦

Cast of Characters

Managers at Time of Crash of *Columbia*

Ron D. Dittemore, Shuttle Program Manager
Linda Ham, Chair of the Mission Management Team
James M. Heflin, Jr. NASA Mission Operations Chief Flight Director
Sean O'Keefe, Administrator of NASA
Gen. Michael C. Kostelnik, Deputy Associate Administrator for the
    Space Shuttle and International Space Station
LeRoy E. Cain, Flight Director of Columbia

Columbia Astronauts (as given in CAIB)

Rick D. Husband, Commander
William C. McCool, Pilot
Michael P. Anderson, Payload Commander
David M. Brown, Mission Specialist
Kalpana Chawla, Mission Specialist
Laurel Blair Salton Clark, Mission Specialist
Ilan Ramon, Payload Specialist

Debris Search Pilots (Killed in Action)

Jules F. Mier, Jr., Debris Search Pilot
Charles Krenek, Debris Search Aviation Specialist

Columbia Accident Investigation Board (CAIB)

Harold W. Gehman, Jr., Chairman
    Admiral, U. S. Navy (retired)
John L. Barry, Major General, U.S. Air Force
Duane W. Deal, Brigadier General, U.S. Air Force

James N. Hallock, Ph.D., Manager, Aviation Safety Division, DOT/ RSPA Volpe Center

Kenneth W. Hess, Major General, U.S. Air Force

G. Scott Hubbard, Director, NASA Ames Research Center

John M. Logsdon, Ph.D., Professor, George Washington University

Douglas D. Osheroff, Ph. D., Professor, Stanford University

Sally K. Ride, Ph.D., Professor, University of California at San Diego

Roger E. Tetrault, Chairman and CEO, McDermott International (retired)

Stephen A. Turcotte, Rear Admiral, U.S. Navy

Steven B. Wallace, Director, FAA Office of Accident Investigation

Sheila E. Widnall, Ph.D., Professor, Massachusetts Institute of Technology ✦

Glossary of Acronyms and Technical Terms

Arsenal Concept The practice of the U.S. Army in which the R&D stages of developing weapons systems are done in-house. The Air Force, by contrast, employs contractors to do the R&D stages as well as the actual production of the weapons systems.

ASTTO The Aerospace Transportation Technology Office, a unit in NASA which works to develop future space vehicles.

Asymmetric Boundary Layer Transition A rough surface on an Orbiter wing capable of producing turbulence.

Anomaly NASA's term for an unexpected occurrence during a mission, also known as a mission contingency.

Apollo A U.S. space flight program that put astronauts on the moon and safely returned them to Earth.

Apollo Applications Program After the trips to the moon, NASA developed new technologies such as Skylab, an orbiting workshop, and the Apollo Telescope Mount.

ASRS NASA's Aviation Safety Reporting System, described in the Preface of this book.

Automatic Responsibility A communication practice at MSFC during the Apollo Program in which every individual who perceives a technical problem within his or her area of expertise automatically assumes responsibility for the problem; if the person lacks the expertise he or she automatically assumed responsibility to communicate the problem up the line so that it could receive the proper technical attention.

Bipod A two-legged brace that attaches the nose of the Orbiter to the external tank.

Bolt Catchers When the solid rocket boosters separate from the external tank the bolts holding them are exploded by pyrotechnics. The bolt catchers are the size of two stacked cans; their purpose is to catch the exploded bolts.

CAIB The Columbia Accident Investigation Board. They issued a report in August of 2003.

CEO Acronym for Chief Executive Officer, the head of a corporation responsible for overall management.

CFO Chief Financial Officer of a corporation.

Challenger The Orbiter or space shuttle that exploded during its launch in 1986, killing seven astronauts.

Columbia The Orbiter or space shuttle that disintegrated during re-entry in 2003, killing seven astronauts.

Communication Transgressions Violations of the basic principle of cooperative society and organization, i.e., telling the truth. They include lying, encouraging others to lie, power plays in the form of bribes, failure to listen to employees, plausible deniability, *ignorantia affectata* (pretended ignorance of illicit acts), the fear of speaking truth to power, suppression of voice, the absence of feedback loops to prevent illicit actions, and the absence of symmetrical information required by capitalism.

Concertive Control The process by which teams or work groups democratically reach decisions by applying accepted decision premises in intense, oral discourse. Such groups usually experience a high degree of organizational identification. Other, less powerful methods of control are simple control—supervision by a boss or owner, technical control—in which assembly lines and computer monitoring supplement simple control, and bureaucratic control—in which rules and regulations supplement simple control. Concertive control increases the total amount of control in the system by virtue of the fact that everyone is a supervisor. The movement from simple control to concertive control is from obtrusiveness to unobtrusiveness.

DAT Debris Assessment Team, the group of engineers who requested additional imagery of *Columbia* in order to determine what effect, if any, the foam debris had.

Deduction A form of reasoning which begins with what is known and then branches out into possible conclusions.

Discursive Closure   The opposite of openness and candor in organizational communication, particularly in upward-directed communication.

ET   The External Tank, the largest component of the space shuttle "stack," contains up to 143,351 gallons of liquid oxygen and 385,265 gallons of liquid hydrogen. The temperatures of the liquid fuels are, respectively, minus 297 degrees and minus 423 degrees Fahrenheit, necessitating a coating of foam insulation. The ET is jettisoned after its fuels are spent, and burns up as it falls toward the Earth.

EVA   Extra Vehicular Activity by astronauts, popularly known as a space walk.

Faster, Better, Cheaper   The slogan Administrator Daniel Goldin brought to NASA during the 1990s. The concept was not supposed to be applied to manned space flight. The results were disappointing, seeming to indicate that faster and cheaper can compromise quality or reliability.

Fault Tree   A graphic device for reverse engineering from a fault to one or more causes. Every possible scenario is diagrammed. The branches of the tree must be eliminated or closed. A branch or branches not closed are the most likely causes of the fault.

Flight Readiness Reviews   Formal discussions, involving charts, of whether all preparations for a launch have been completed, that rationales for possible anomalies have been approved.

Formal Versus Informal Organization   The formal organization is depicted by lines and boxes of an organizational chart. The boxes are positions and units, the lines are the communication channels that connect them, indicating which units report to whom, and which ones have authority to direct or give orders to others. The informal organization, by contrast, is personal in nature, consisting of unofficial relationships developed in car pools and other chance meetings.

Groupthink   A mode of thinking when the members of a group have a high motivation to achieve agreement. That striving for unanimity overrides the ability to realistically evaluate alternative courses of action.

Hierarchy   The graded, vertical series of status levels in an organization. It is usually thought of as formal, that is, specified and acknowledged; it can be informal as well.

Hindsight Bias This involves using knowledge of outcomes to reconstruct the antecedents of the outcomes. If, for example, a machine operates successfully once or twice despite a faulty part, one assumes that everything will continue to work well.

Hypergolic Fuels Fuels that ignite spontaneously when mixed together. During the fueling of *Columbia* some of these materials were spilled on the Orbiter's left wing.

Ideal Managerial Climate The characteristics of this theoretical model are supportiveness; participative decision making; trust, confidence, and credibility; openness and candor; and an emphasis on high performance goals.

Inconel A material used in the fittings on the leading edge of the Orbiter's wings.

Induction A form of reasoning in which facts, evidence, or data are brought to bear on hypotheses in both positive or negative ways.

ISS International Space Station, a cooperative venture involving several nations. It is serviced by the U.S. Orbiter and by Russian spacecraft. An Orbiter was scheduled to haul the structural skeleton of a new "node," or addition, to the ISS in February of 2004.

JSC The Johnson Space Center, a NASA field center near Houston, Texas, where Orbiter missions are monitored and managed.

Kapton Wiring The space age wiring used in the Orbiter.

KSC Kennedy Space Center at Cape Canaveral, Florida, a NASA field center where Orbiters are launched and land.

Matrix Management The method of managing an organization of relatively permanent substantive departments with project offices that come and go. The lines of communication are vertical, horizontal, and diagonal, thus resembling a matrix.

Micrometeoroids Tiny meteoroids that can damage the Orbiter.

Mishap Interagency Investigation Board NASA's "independent" investigating body set up after the *Challenger* accident. After the *Columbia* accident it would be expanded and transformed into CAIB.

Monday Notes A communication practice at MSFC during the Apollo Program in which openness and candor was achieved in a highly disciplined manner. The practice is explained in detail in Chapter Four of this book.

MSFC The Marshall Space Flight Center, a NASA field center in Huntsville, Alabama.

Mission Management Team The group of managers responsible for each launch, e.g., STS-107.

NASA The (U.S.) National Aeronautics and Space Administration.

NOAA The (U.S.) National Oceanic and Atmospheric Administration.

Normalization of Deviance A sociological concept specifying a gradual acclimatization to, or acceptance of, unexpected and undesired results.

NSF National Science Foundation.

Orbital Debris Risks The threats to spacecraft from manmade "junk" orbiting around the Earth left over from previous launches.

Orbiter Popularly known as the space shuttle, the Orbiter is about the size of a commercial airliner; it normally carries a crew of seven, including a Commander, a Pilot, and five Mission or Payload Specialists.

Organizational Communication The study of organizations as communication systems, with messages sent (and not sent) in all directions, vertically, diagonally, horizontally. Communication constitutes organizations in the sense that organizations cannot come into existence and maintain their existence in the absence of communication. It is often assumed that there is a correlation between effective communication and organizational success, and between ineffective communication and organizational failure. The study of such phenomena can be done empirically, interpretatively, and critically.

Organizational Culture The rough draft of individual behavior within an organization. Individuals are suspended in webs of significance or meaning of their own making. The "stronger" the culture, the more the behaviors of employees will be similar. The weaker the culture, the more they will be different and less predictable. Language, symbolism, and meaning are the heart of culture.

Organizational Identification A person identifies with an organization when she or he seeks to select the alternative that best fits the perceived interests of the organization. The higher degree of identification the more the individual's identity is shaped by membership in the group, and the more the person develops a "we-self." Identification is a process while a related concept, organizational commitment, is more behavioral.

O-Rings   The large gaskets that seal field joints on the solid rocket boosters, to prevent searing gases from escaping. One of these failed in 1986, causing the explosion of *Challenger.*

OSHA   The Occupational Safety and Health Administration, a U.S. government office responsible for monitoring risks to health and safety in the workplace.

OSP   Orbital Space Plane, a vehicle NASA desires to build, which would reach the International Space Station faster and cheaper than the Orbiter.

OTC   The Original Technical Culture of NASA during its glory days of the 1960s. The categories include research and testing, in-house technical capability, hands-on experience, exceptional people, risk and failure, frontier mentality, open communication, and a high degree of organizational identification.

Penetration   A communication practice in which the organization buying a product from another penetrates or gains access to the selling organization in order to anticipate problems with the product. At MSFC in the Apollo Program there was an internal, vertical application of the practice.

Plasma   A gas so hot it glows with electricity. It can slice through steel as if it were butter.

Plausible Deniability   A communication transgression in which leaders of an illicit conspiracy take care to be sure they can persuasively deny responsibility for antisocial acts of others.

Police Procedural   A fictional literary form which begins with a crime—that much is known. Then the reader learns of the evidence as the police gather it, as they develop and reject false leads, and as they finally arrive at the solution.

Problem Reporting and Corrective Action System   NASA's method of attempting to track thousands of problems with the space shuttle and assure that they receive the proper attention and a technical solution.

Pyrotechnics   Explosive devices used to separate components of the stack, and to blow off the landing gear doors of the Orbiter in case they don't open for a landing.

Rationalization   The process by which bureaucracy increases its control of society and its organizations. Instrumental rationality calculates ends and criteria as well as means. Value rationality assumes that ultimate values such as the Golden Rule should serve as the major premise or criterion for decisions.

RCC Reinforced Carbon-Carbon, the material used to keep the wings and nose cap of the Orbiter from burning up from plasma during re-entry to the Earth's surface. It is a sheet of carbon which has been immersed in carbon resin.

R&D Research and Development, the process of conceiving, designing, fabricating, and testing hardware. It is to be distinguished from the operational stage in which a spacecraft is treated as a certified vehicle of high reliability, e.g., a commercial airliner.

Redundancy More information than is logically or ideally necessary—e.g., repetition and restatement. An organization can also have redundant channels to better assure crucial information gets up and down the line. The word is also used in a technomorphic sense, as when rockets and spacecraft have backup systems in case something fails.

RIF Acronym for Reduction in Force, the layoff of U.S. Government employees.

Rogers Commission The popular name of the Presidential Commission on the Space Shuttle Challenger Accident. The Commission was chaired by William P. Rogers, former Secretary of State, and issued its report in June of 1986.

Saturn V The moon rocket developed at MSFC for the Apollo Program.

SEC The Securities and Exchange Committee, a regulatory agency responsible for the stock market.

Semantic-information distance This refers to the degree of understanding-minunderstanding among people at different levels of an organization, as well as the degree of knowledge-ignorance of ideas crucial to the organization's purpose and rules.

Semantogenic Miscommunication The idea that the common meaning of a word can lead an individual or group to misunderstand or misinterpret reality.

Shuttle Stack The solid rocket boosters are bolted to the mobile launch platform tightly enough to bear the weight of the rest of the components. The external tank is attached to the solid rocket boosters, and the Orbiter is then attached to the ET at three points, two at the bottom and a bipod attachment near the nose.

Solid Rocket Boosters Two solid-fuel tanks with rocket engines, attached to the external tank, which provide 85 percent of the thrust during the launch of the Orbiter. After they are jettisoned, parachutes slow their descent into the ocean where they are recovered and reused. A faulty O-Ring, or giant washer, allowed hot gases to escape from a field joint during the launch of *Challenger,* causing its fatal explosion.

Sputnik 1 The first artificial satellite to orbit the Earth, launched by the Soviet Union in 1957. This achievement intensified the space race between the Soviet Union and the United States.

STS-107 The designation for the last flight of *Columbia* launched on January 16, 2003.

Thermal Protection System Three different insulating components of the Orbiter, designed to keep the spacecraft from burning up from the temperatures up to 3,000 degrees Fahrenheit experienced during re-entry. The insulating components are tiles, blankets, and reinforced carbon-carbon (RCC). RCC is too heavy to be used applied to the entire Orbiter; it is placed on the leading edges of the wings and the nose cap of the spacecraft.

Topoi A Greek term for places to look for arguments. The three *topoi* for rocket research and development are reliability or safety, time or schedule, and cost. Difficult decisions often involve tradeoffs among the three. "Give me three more days and $10,000 and I'll increase the reliability of the system," is a hypothetical example of a tradeoff. The singular form of *topoi* is *topos*.

United Space Alliance A consortium formed by two aerospace contractors, Boeing and Lockheed Martin, given responsibility for day-to-day operations of the shuttle.

USSTRAT U.S. Strategic Command. A unit in the Defense Department, USSTRAT had the capability to provide imagery of *Columbia* in orbit by means of telescopes on land and in spy satellites.

V-2 A military missile developed in World War II by the German Rocket Team Wernher von Braun at Rocket Center Pennemünde.

Voice A metaphor for the degree to which employees participate and have influence in organizational decision making, the degree to which their views are listened to by those at the top of the hierarchy.

Whorf Hypothesis Also known as the Sapir-Whorf Hypothesis, this is the linguistic principle that the language of the group unconsciously creates the group's reality.

Zipper Effect The loss of even one insulating tile on the Orbiter could have the effect of stripping away other tiles, making the spacecraft vulnerable to super-hot gases. ✦

References

Barker, J. (1993, September). Tightening the iron cage: Concertive control in self-managing teams. *Administrative Science Quarterly, 38*, 408–437.

Barker, J., and Tompkins, P. (1994). Identification in the self-managing organization: Characteristics of target and tenure. *Human Communication Research, 21*, 223–240.

Barnard, C. (1938). *The functions of the executive*. Cambridge, MA: Harvard University Press.

Broad, W. (1992, March 12). Bush nominates TRW executive to lead space agency to new era. *New York Times*, A1, A9.

Bullis, C. (1991). Communication practices as unobtrusive control: An observational study. *Communication Studies, 42*, 254–271.

Bullis, C., and Tompkins, P. (1989). The Forest Ranger revisited: A study of control practices and identification. *Communication Monographs, 56*, 287–306.

Burke, K. (1935/1954). *Permanence and change: An anatomy of purpose*. Indianapolis, IN: Bobbs-Merrill.

———. (1964). The rhetoric of Hitler's 'Battle.' In S. Hyman (Ed.), *Terms for order* (pp. 95–119). Bloomington, IN: Indiana University Press.

———. (1966). Terministic screens. In K. Burke (Ed.), *Language as symbolic action* (pp. 44–62). Berkeley: University of California Press.

———. (1969). *A rhetoric of motives*. Berkeley: University of California Press. (Originally published in 1950).

———. (1972). *Dramatism and development*. Barre, MA: Clark University Press.

CAIB Report. See *Report of the Columbia Accident Investigation Board*.

Cheney, G. (1987, February). The linkage of sacrifice and purpose in the rhetoric of the *Challenger* disaster: Media accounts and the reinforcement of a national 'mission.' Paper presented at the annual convention of the Western Speech Communication Association, Salt Lake City, UT.

———. (1999). *Values at work: Employee participation meets market pressure at Mondragon*. Ithaca, NY: Cornell University Press.

Cheney, G., Christensen, L., Zorn, T., and Ganesh, S. (2004). *Organizational communication in an age of globalization: Issues, reflections, practices.* Prospect Heights, Ill: Waveland Press, Inc.

Cheney, G., and Tompkins, P. (1987). Coming to terms with organizational identification and commitment. *Central States Speech Journal,* 38, 1–15.

Deetz, S. (1992). *Democracy in an age of corporate colonialization.* New York: Hampden Press.

Dunar, A., and Waring, S. (1999). *Power to explore: A history of Marshall Space Flight Center, 1960–1990.* Washington, D.C.: NASA.

Edwards, R. (1978). *Contested terrain: The transformation of the workplace in the twentieth century.* New York: Basic Books.

Ehrenreich, B. (2001). *Nickel and dimed: On (not) getting by.* New York: Henry Holt.

Eisenberg, E., and Riley, P. (2001). Organizational culture. In F. Jablin and L. Putnam (Eds.), *The new handbook of organizational communication* (pp. 291–322). Thousand Oaks, CA: Sage.

Feynman, R. (1986). *'Surely you're joking, Mr. Feynman!'* New York: Bantam Books.

——. (1988, February). An outsider's inside view of the *Challenger* inquiry. *Physics Today,* 26–37.

——. (1989). *'What do you care what other people think?'* New York: Bantam Books.

Gane, N. (2001). *Max Weber and postmodern theory: Rationalization versus reenchantment.* New York: Palgrave.

Geertz, C. (1973). *The interpretation of cultures.* New York: Basic Books.

Gladwell, M. (2002, July 22). The talent myth. *New Yorker* (28–33).

Gossett, L. (2001). *More than guests—less than members: Examining issues of organizational identification within the temporary help industry.* Unpublished Ph.D. dissertation, University of Colorado at Boulder.

Hamel, G. (2000). *Leading the revolution.* Boston: Harvard Business School Press.

Hirschman, A. (1970). *Exit, voice, and loyalty; responses to declines in firms, organizations, and states.* Cambridge, MA: Harvard University Press.

Jablin, F. (1979). Super-subordinate communication: The state of the art. *Psychological Bulletin,* 86, 1201–1222.

——. (1985). Task/work relationships: A life-span perspective. In M. Knapp and G. Miller (Eds.), *Handbook of interpersonal communication* (pp. 615–654). Newbury Park, CA: Sage.

Jensen, C. (1996). *No downlink: A dramatic narrative about the* Challenger *accident and our time.* New York: Farrar, Straus, Giroux.

Johnson, S. (2002). *The secret of Apollo: Systems management in American and European space programs.* Baltimore: Johns Hopkins Press.

Keneally, T. (2002, June 17 and 24). Cold sanctuary: How the Church lost its mission. *New Yorker,* 58, 60, 62–66.

Langewiesche, W. (2003, November). *Columbia's* last flight: The inside story of the investigation—and the catastrophe it laid bare. *Atlantic Monthly,* 58ff.

Larson, G., and Tompkins, P. (2003). Ambivalence and resistance: A study of Management in a concertive control system. Paper presented at the National Communication Association Convention, Miami, FL.

Levitt, A. (2002, October 14). Interview with A. Levitt. *Fortune,* 64.

Mailer, N. (1971). *Of a fire on the moon.* New York: New American Library.

McCurdy, H. (1993). *Inside NASA: High technology and organizational change in the U.S. space program.* Baltimore: Johns Hopkins University Press.

———. (2001). *Faster, better, cheaper: Low-cost innovation in the U.S. space program.* Baltimore: Johns Hopkins University Press.

Messerschmidt, J. (1995). Managing to kill: Masculinities and the space shuttle *Challenger* explosion. *Masculinities,* 3, 1–22.

Michaels, E., Handfield-Jones, H., and Axelrod, B. (2001). *The war for talent.* Boston: Harvard Business School Press.

Modaff, D., and S. DeWine (2002). *Organizational communication: Foundations, challenges, and misunderstandings.* Los Angeles, CA: Roxbury Publishing Company.

Nieburg, H. L. (1966). *In the name of science.* Chicago: Quadrangle Books.

Pasternak, J. (1992, August 5). When the sky isn't the limit. *Los Angeles Times,* A1, A12, A13.

Perrow, C. (1984). *Normal accidents: Living with high-risk technologies.* New York: Basic Books.

Phillips, K. (2002). *Wealth and democracy.* New York: Broadway Books.

Rachels, J. (1993). *The elements of moral philosophy.* New York: McGraw-Hill.

Redding, W. C. (1972). *Communication within the organization.* New York: Industrial Communication Council.

———. (1988). Rocking boats, blowing whistles, and teaching speech communication. *Communication Education,* 245–58.

Report of the Columbia Accident Investigation Board, Volume I (August, 2003). CAIB.

Report of the Presidential Commission on the Space Shuttle Challenger Accident (June 6, 1986). The Rogers Commission.

Starbuck, W., and Milliken, F. (1988). *Challenger:* Fine-tuning the odds until something breaks. *Journal of Management Studies,* 25, 319–40.

Tillich, P. (2000). *The courage to be.* New Haven: Yale University Press. (Originally published 1952.)

Tompkins, P. (1968). Organizational communication: A state-of-the-art review. In G. Richetto (Ed.), *Conference on organizational communication* (pp. 4–26). Huntsville, AL: MSFC/NASA.

———. (1977). Management qua communication in rocket research and development. *Communication Monographs,* 44, 1–26.

———. (1978). Organizational metamorphosis in space research and development. *Communication Monographs,* 45, 110–118.

———. (1987). Translating organizational theory: Symbolism over substance. In F. Jablin, L. Putnam, K. Roberts, and L. Porter (Eds.), *Handbook of organizational communication: An interdisciplinary perspective.* Newbury Park, CA: Sage.

———. (1990). On risk communication as interorganizational control: The case of the Aviation Safety Reporting System. In A. Kirby (Ed.), *Nothing to fear: Risks and hazards in American society.* Tucson: University of Arizona Press.

———. (1991). Organizational communication and technological risk. In Lee Wilkins and Philip Patterson (Eds.), *Risky business: Communicating issues of science, risk, and public policy* (pp. 113–129). New York: Greenwood Press.

———. (1993). *Organizational communication imperatives: Lessons of the space program.* Los Angeles: Roxbury Publishing Company.

Tompkins, P., and Anderson, E. (1971). *Communication crisis at Kent State: A case study.* New York: Gordon & Breach.

Tompkins, P. and Cheney, G. (1983). Account analysis of organizations: Decision making and identification. In L. Putnam and M. Pacanowsky (Eds.), *Communication and organizations: An interpretive approach* (pp. 179–210). Newbury Park, CA: Sage.

———. (1985). Communication and unobtrusive control in contemporary organizations. In R. McPhee and P. K. Tompkins (Eds.), *Organizational communication: Traditional themes and new directions* (pp. 179–210). Newbury Park, CA: Sage.

Tompkins, P., Fisher, J., Infante, D., and Tompkins, E. (1975). Kenneth Burke and the inherent characteristics of formal organizations: A field study. *Speech Monographs,* 42, 135–142.

Tompkins, P., Heppard, K., and Melville, C. (1998). Deviance from normality or Normalization of deviance? Making sense of the *Challenger* launch decision. *Organization,* 5, 620–629.

Vaughan, D. (1996). *The Challenger launch decision: Risky technology, culture, and deviance at NASA.* Chicago, IL: University of Chicago Press.

Weber, M. (1978). *Economy and society.* G. Roth and C. Wittich (Eds.). Berkeley: University of California Press.

Weick, K., and Ashford, S. (2001). Learning in organizations. In F. Jablin and L. Putnam (Eds.), *The new handbook of organizational communication: Advances in theory, research, and methods.* Thousand Oaks, CA: Sage.

Whorf, B. (1956). *Language, thought and reality.* Published jointly by the Technology Press of M.I.T. and John Wiley & Sons, New York.

Wilford, J. N. (1969). *We reach the moon.* New York: Bantam Books. ✦

Name Index

A, B

Adwan, T. 155
Alexander, Sonja 24
Anderson, John 8
Anderson, Michael P. 14
Anderson, Nick 236
Anderson, William C. 140
Aristotle 49, 104, 108
Armstrong, Neil xvii, 102–103, 112
Ashford, Susan J. 85, 105
Austin, Lambert 145
Axelrod, B. 230–231
Baker, Daniel 135
Balch, J. M. 81
Barker, James 108, 194, 221
Barnard, Chester x, 71, 74–75, 204
Barstow, David 23, 129, 134
Bass, Lance 33
Bearden, David 177–178
Behnken, Robert L. 146–147
Bennett, James 20
Benz, Frank 149
Berra, Yogi 156–157
Blakemore, Bill 9
Bloss, Laura 152
Boisjoly, Roger 39, 124, 136
Bottos, Lynda 148
Bowersox, Kenneth D. 21
Bridges, Roy D. 157

Broad, William J. 17, 24, 32, 37, 42, 132, 140
Broder, John M. 26, 38, 130, 131–132
Brokaw, Tom 9
Brown, Aaron 9
Brown, David M. 14
Brown, Laura J. 143
Brownback, Sam 236
Brownlee, Donald 31
Budarin, Nokolai M. 21
Bullis, Connie 107
Burke, Kenneth 48, 104, 106–108, 111, 151–152, 179
Bush, H. W. 175–176
Bush, President George W., Jr. 10, 13, 26, 34, 234–235, 253
Bush, Jeb 43

C, D

Cabana, Robert D. 21, 22
Cabbage, Michael 249
Cain, LeRoy E. 133–134, 155, 238–239
Camera, Frankie 12
Campbell, Carlisle C. 140, 238
Canedy, Dana 12
Card, Andrew H., Jr. 10
Carr, David 21

Subject Index

Note: [G] indicates term defined in Glossary.

Manufacturing Engineering Lab 72

Mariner 9, 64

Marshall Space Flight Center *see* MSFC

Marxism 215

Massachusetts Institute of Technology (MIT) 24, 42, 133, 134

Matrix management [G] 60, 72, 182, 187, 250

McDermott International 42

McKinsey and Company 222

McWane Corporation 205–209

Mercury 63, 64

Mercury Program *see Mercury*

Michoud Assembly Facility 33–34, 37, 70, 81

Michigan State University xx, xxi, 4, 154

Micrometeoroids [G] 131, 166

MIIB, Mishap Interagency Investigation Board [G] 32

Milestones 91; *see also* PERT

Miscommunication xviii, xix, 3, 4, 171, 178, 180, 181, 219, 233; *see also* Communication barriers, blockages

Mishap Interagency Investigation Board 22, 32, 163

Missile Command (U. S. Army) xiii

Mission contingency 16

Mission Control Center 55

Mission Evaluation Room 172, 173, 239

Mitosis 71–82

MMACS: Maintenance, mechanical and crew systems 140

MMT, Mission Management Team [G] 172–175, 185, 197

Monday Notes [G] 80, 82–87, 105, 116–117, 190, 193

Moon rocket 56, 63, 96; *see also* Saturn V

Morton-Thiokol 37, 39, 113, 117–118, 120, 123–125, 135, 194

MSFC [G] xii, xvii, 3, 55, 56, 58, 59, 60, 67–109, 115, 117, 124, 171, 190, 193, 195, 202, 204, 249

MTF: Mississippi Test Facility xiv, 70, 81

Mushroom problem 81–82

NACA: National Advisory Committee for Aeronautics 52, 55, 57

NASA: The (U.S.) National Aeronautics and Space Administration [G] xi–xx, xi–xv, 1–18, 19–41, 47–64, 67–108, 111–128, 129–162, 163–199, 201–230, 233–255

National Imagery and Mapping Agency 148–149

National Science Foundation *see* NSF

Navy 54, 151, 163, 180, 236

Nazi(s) 52, 75–76, 108, 151

New Space Transportation System 196

New York Times [Premium Archives] xxiii–xxiv, 1, 26, 43, 129

NOAA [G] 10

Normal accidents 121, 123

Normalization of deviance, normalized deviance [G] 63, 119, 122–127, 180–181

Nose cone 39

NSF [G] xix

Nuremberg trials 181

O, P

Occupational Safety and Health Administration *see* OSHA

Ontological anxiety 254–256

Open communication 83–87, 102–105, 123, 189, 203

Open loops *see* Open communication

Orbital Debris Risks [G] 131, 166